HOW DO YOU GET AN EGG INTO A BOTTLE?

THIS IS A CARLTON BOOK

First published in Great Britain in 2019 by Carlton Books Limited
an imprint of the Carlton Publishing Group
20 Mortimer Street
London W1T 3JW

10 9 8 7 6 5 4 3 2 1

A catalogue record for this book is available from the British Library

ISBN 978-1-78739-272-4

Printed in Dubai

Author's acknowledgments:
The laws of nature are immutable. They are, however, not recorded in any
statute books that might be available to humanity to study the rules that govern
our cosmos. So, in a laborious process going back to the dawn of history,
scientific man has tried to fathom the mysteries of the universe. These collective
efforts, spanning millennia, provide a rich source of challenging conundrums.
In compiling this book, we enjoyed permission from other authors to use their
material, including Scot Morris, Stephen Barr, and the Encyclopedia Britannica.
Thanks should go to Sarah Gerrard for her patient support, and to Bob Anthony
on whom we tested some of our ideas.

Content in this book has previously been published in *How Do You Get An Egg
Into A Bottle?* and *How Do You Walk On Fire?*

The publishers would like to thank the following sources for their kind
permission to reproduce the pictures in this book.

Carlton Books/Peter Liddiard; NASA; Shutterstock; Wikimedia Commons

Every effort has been made to acknowledge correctly and contact the source
and/or copyright holder of each picture and Carlton Books Limited apologises
for any unintentional errors or omissions, which will be corrected in future
editions of this book.

ERWIN BRECHER & MIKE GERRARD

HOW DO YOU GET AN EGG INTO A BOTTLE?

SCIENTIFIC PUZZLES TO BAFFLE AND BEMUSE YOUR BRAIN

CARLTON
BOOKS

Contents

Introduction

This book is for the curious.

We are surrounded by phenomena often dressed in everyday clothes which are rarely recognized as such until, at odd moments, we stop to ask ourselves: What arcane rules make the physical world behave as it does?

Whereas philosophy is concerned with the "why" in probing the innermost corners of spiritual existence, science explores the "how" by processes of inductive and deductive reasoning. The former process is an exercise in arriving at general hypotheses and theories from specific observations and experiments, while the latter reverses the procedure by starting with the theory to show that it will satisfactorily explain experimental results. This dichotomous approach is at the core of the scientific method, which was first articulated during the Renaissance but which in fact has been used throughout human history.

More often than not we take the workings of the physical world around us for granted without realizing the profound effect these, many apparently mundane, phenomena have on our understanding of the universe.

Sir Isaac Newton (1643–1727), one of the great scientists of all time, discovered the principles of universal gravitation following the fall of an apple. The story might be anecdotal, but Newton's achievements are monumental.

The time is New Year's Eve, 1670. Picture yourself sitting at home and seeing an object fall. Would you have given it a

"All men by nature desire knowledge"

Aristotle (384–322 BCE)

second thought? Probably not. But if you had, you would most likely have said to yourself: "What else could it do but fall to the floor?" In other words, you would have considered it axiomatic that everything falls down and not up. However there are no axioms in physics; that is to say, there are no basic principles that are assumed to be true without proof or inherent logic, as is frequently the case in geometry.

On the contrary, there is not a single aspect of our physical world which is not investigated, analyzed and evaluated by scientists who are determined to find an explanation for all phenomena, be they esoteric or mundane. The relevant literature is enormous, with some questions remaining unresolved while opinions on other theories differ.

We can't all be Newtons, but we can have an inquiring mind and derive a great deal of satisfaction from understanding what makes things "tick". In some cases we shall find the answer, using basic principles of physics and a modicum of problem-solving ability. With others the answer will make immediate sense, with a tinge of regret for not having thought of the explanation in the first place. Finally there will be quite a few where the solution will be a revelation.

In all cases, the intelligent reader will close this book feeling that his horizon has widened beyond expectation.

Erwin Brecher

Introduction

I have always loved puzzles of all sorts, from crosswords to detective novels. There is something immensely satisfying about tracking down the solution from a few apparently unconnected

bits of information. I think that is why I have always been fascinated by science. The scientific method of proposing hypotheses and then checking them out by experimentation is immensely attractive.

But solving a scientific conundrum has an additional attraction that no thriller writer can hope to match. Having worked out that "the butler must have done it" is satisfying, but at the end of the day it is just a story. Solving a scientific puzzle also teaches you something about the way the universe works.

It must be one of the most exciting and creative of human activities to discover something absolutely new that no one has ever known before. The Einsteins, Newtons, Darwins, and so on, must all have shared an incredible sense of elation having made that final connection in their minds.

This book does not pretend to be in quite the same league, but it is hoped that you can experience a similar sensation. You may find an observation that you have never seen for yourself, or a question that you have never thought to ask.

The solution to it lies in your own head somewhere; all you need do is formulate your own thoughts to produce the solution. The process can be completely absorbing; it is where the stories of absent-minded professors originate. Archimedes was so involved with his own problem that he foolishly asked an invading Roman soldier to get out of the light, and got himself killed.

To get the most out of this book you are advised not to rush between the problems and the stated solutions; that will spoil the fun! Think about things. It may take several days or a little research to decide on what you think the answer is. Only then look at our solution. You may even disagree with what we say – your solution may be superior to ours. Happy solving!

Mike Gerrard

Sight and Sound

Chapter 1

How can I turn red to green?

A light-emitting diode (LED) is a small solid-state device that gives off light when a current is passed through it. It is commonly used in clocks to give a clear red, or sometimes green, digital display.

There is a type of LED on the market that glows red when the current is passed one way through it, and green when the current flows the other way. An alternating current flows rapidly one way and then the other.

What would this LED look like if it were connected to an alternating source?

How do I dull my headlights?

Many night driving accidents are caused by the glare of headlights from approaching cars. My friend Jonathan, who fancies himself an inventor, came up with an idea offering an effective and inexpensive solution to the problem.

"Let us introduce polarizing filters in front of headlights," he said, "to polarize light horizontally, which would absorb all photons whose electric vectors are vertical. Conversely, let us use windshields with a filter turned 90° to the first, absorbing light emitted by the headlights." On the face of it, this would be the perfect solution, as the light from the approaching car would be blocked out, while all other objects would be visible.

Would this idea work? If so, why hasn't it been adopted?

Q03
Sight and Sound

Can a mirror turn me upside down?

It is often said that mirrors reverse left-right but not up-down. Can you think how a single plane mirror can:

1. **reverse up-down as well?**
2. **reverse up-left and down-right?**
3. **reverse up-down but not left-right?**

Q04
Sight and Sound

So near and yet so far?

Hans Lippershey (1570-1619) was a German-Dutch spectacle-maker who is credited with the invention of the telescope; he later adapted it to create binoculars. Unable to patent the invention because of the ease with which it could be copied, he was nevertheless granted 900 florins for the design. His legacy lives on – two planets and a lunar crater bear his name.

We are very used to the idea of telescopes and binoculars, but what are they doing?

Do they make distant objects look closer or bigger?

What about a magnifying glass: does it make things look bigger, or is it doing something else?

Solution on page 182-3

Why isn't the whole sky as bright as sunlight?

Heinrich Wilhelm Matthaeus Olbers (1758–1840), a German astronomer, hypothesized that if the universe were infinite and contained an infinite number of stars, then the whole sky should be as bright as sunlight. After all, the observable universe alone indicates that the Milky Way is only one of several hundred thousand million galaxies, with each galaxy containing in turn some hundred thousand million stars. It seems therefore inconceivable that any line of sight could miss a star.

You can compare it with a blackboard on which you are asked to make a mere one million chalk marks. Surely the blackboard would appear white? Why then do we not see the sky brightly lit at night?

Glaringly obvious?

Ordinary sunglasses work by absorbing some of the light that would otherwise be transmitted. Polarizing sunglasses work in a slightly different way: they will transmit only light that is vibrating vertically. As light reflected from wet roads, water, and so on tends to vibrate horizontally, this cuts down glare from such surfaces. If you are wearing a pair of polarizing sunglasses and hold up another polarized pair rotated through 90 degrees, the lenses appear black because no light can now pass through both sets of glasses.

Could you insert a third set of polarizing sunglasses between the first two in such a way as to permit some light to be transmitted?

Q07
Sight and Sound

Transparent object

What transparent object, found in many households, becomes less transparent when wiped with a clean, dry cloth?

Q08
Sight and Sound

Can you travel faster than light?

The other day, I was trying to explain relativity theory to a friend. Einstein had predicted, and experiments had confirmed, that as objects moved faster they also increased in mass. At the speed of light, any object would have infinite mass and so would need infinite energy to make it go faster.

"This means that there is a maximum speed at which anything can travel, and only massless things like light can travel at that speed," I explained. My friend is very bright, and after a little thought he came up with three situations that seemed to contradict Einstein.

Is any of them valid?

1. The poles outside barber shops have helical stripes painted round them. If the pole is rotated, its stripes are seen to travel along it. If the rotation is fast enough the stripes should travel faster than light.
2. If you shine a torch on a wall, you will see a spot of light. You can make this spot move along the wall by rotating the flashlight. The speed of the spot depends on the distance from the wall and the speed of rotation of the torch. If you shine a laser on a target some miles away, the spot of light formed could be made to move faster than the speed of light by rotating the laser.
3. Very small particles, such as neutrons, can penetrate matter very easily. Imagine such a particle, travelling at nearly the speed of light, entering a block of glass. The particle is so small that it will not be impeded by the glass, but light is slowed down to about two-thirds of its speed in air. Therefore, the particle would be travelling faster in the glass than the light.

Solution on page 183

The spectrum

One of the first art lessons learned at school is that the colour green can be obtained by mixing yellow and blue paint. On the other hand, strangely enough, if you project yellow and blue light, through appropriate gelatin sheets, onto a screen, you obtain white light.

Can you think of an explanation?

Solution on page 184

Q10
Sight and Sound

Why can't I tune my radio?

Nobel prize-winning Italian inventor and engineer **Guglielmo Marconi (1874-1937)** is known as the inventor of radio – he successfully broadcast the first trans-atlantic radio signal. His Marconi Company radios, which were operated by "Marconi Men", were instrumental in numerous significant events, including sending the distress signals that saved the RMS Titanic survivors. During his Nobel Prize acceptance speech however, he freely admitted that he didn't quite understand how it all worked.

On a recent drive around the country, I was listening to a very interesting programme on the car radio. As I moved away from a transmitter, the reception got progressively worse, so I had to retune the radio to pick up the same station on a nearer transmitter. I had to repeat this process several times during the journey. This really affected my enjoyment of the programme.

Why do they not ensure that all transmitters broadcasting the same station use the same frequency?

Solution on page 184

Q11
Sight and Sound

Can you see 3D?

There have been a number of attempts to create a three-dimensional effect from two-dimensional images. One of the earliest was a device called a stereoscope, through which one looked at specially printed cards. Much later, "3D" came to the cinema screen. Through red- and green-lensed spectacles passed out upon entering the theatre, cinema-goers enjoyed an effect that was said to be so realistic that a train speeding out of the screen towards the audience created panic. This experiment was soon abandoned as the 3D glasses needed found little favour with movie audiences.

Since then, more sophisticated techniques have been devised, leading to specially designed patterns (stereograms), usually printed in colour, that are being marketed under various trade names. Once the viewer has mastered the technique, the effect is spectacular.

It is, of course, one thing to enjoy the images, but quite another to understand how it works. Can you find an explanation for the 3D effect of the patterns, and prove that your theory is correct?

Q12
Sight and Sound

Black to the future

Our modern world is full of light. Walk through any city centre at night and you will see all sorts of coloured lights, from white through every colour in the rainbow.

But it occurs to me that there is one colour that is omitted, and that is black. This seems a great pity, because I could think of many uses. For example, one could use a black bulb in a photographic darkroom, avoiding the necessity of costly blackout material. During a war, searchlights are used to illuminate bombers at night and also to dazzle the pilots. A black searchlight could be used during the day to plunge the flyers into darkness.

Anything wrong with this scheme?

Q13
Sight and Sound

Can you stop an echo?

When something makes a loud bang between us and the side of a house or a cliff, we hear it and its echo. What would we hear if the source of sound were touching the surface?

Solutions on page 185

Q14
Sight and Sound

Rays of the Sun

When conditions are right, the Sun's rays streak across the sky from behind a distant cloud or mountain. Meteorologists call these crepuscular rays. They always fan out from a point that seems to be just behind the obstruction.

But wait, aren't the Sun's rays supposed to be parallel when they reach Earth?

How does a cloud or mountain cause the rays to diverge that way?

Solution on page 185

The well-trained sunglasses

You will come across many puzzles in science to which no amount of reasoning will provide the answer if you do not happen to be a specialist in that particular field. Take, for instance, photosensitive sunglasses. It is not too far-fetched to recognize that sunlight can cause a physio-chemical reaction that will darken certain substances. After all, your skin tans on exposure to the Sun. But tanning takes time. Furthermore, your skin does not turn pale again the moment you retreat into shade.

Can you at least speculate as to the principle that might be involved in the case of our well-trained sunglasses?

How do I blow up my stereo?

When setting up my new hi-fi system, I noticed that the instructions were very emphatic about connecting the wires to both speakers in the same way. In fact, the wires were colour-coded to try to make it difficult to get wrong.

What would happen if the loudspeakers were incorrectly connected?

What's in the middle of a rainbow?

The rainbow is one of nature's most uplifting spectacles. Occurring as it does when sunshine meets suspended water droplets, as after a summer shower, the rainbow is a universal symbol of optimism. We most often think of a rainbow as an arc, round on top with legs below, only because water droplets are rarely seen below the horizon. You can see full-circle rainbows if you stand near a waterfall or a lawn sprinkler, or if your vantage point is "above the weather", as from a cliff or the top of a tall building. When you see a more familiar arc rainbow, imagine extending the arc's curvature into a complete circle. What feature, then, will you see at the circle's centre?

Solution on page 186

How did we avoid rockets?

Those of us in Great Britain who lived through World War II will remember, without much nostalgia, V-1 and V-2 rockets replacing the bombing raids that had become too expensive for the Germans in terms of losses in men and planes.

It is in man's nature to be adaptable. We soon found that the V-1s presented no danger as long as you could hear them. However, as soon as the buzzing of the engine stopped, it was advisable to dive under the bed. No such strategy was available with the V-2.

Why not?

Solution on page 186

Light fantastic

Imagine two blacked-out rooms bathed in yellow light. A piece of white paper examined in both rooms appears to be exactly the same shade of yellow. Another piece of paper appears black in one of the rooms and striped red and green in the other.

How could this be?

Q20
Sight and Sound

While active, **Concorde (1976-2003)** flew just under 50,000 flights and could reach speeds of Mach 2.04, or 1,354 miles per hour. It was one of only two commercial supersonic transports, the other being the Soviet-built Tupolev Tu-144, which flew only 55 commercial flights. It's fastest transatlantic crossing was achieved on 7th February 1996 when it flew from New York to London in just 2 hours 52 minutes and 59 seconds.

Can you fly faster than sound?

Even if you have never flown in Concorde, that ingenious product of Anglo-French technology, you know that this plane was the only supersonic aircraft in commercial use. You also know that Mach 1 stands for the speed of sound.

Now answer the following questions:

1. **Do you hear the sonic boom when, travelling in Concorde, you reach Mach 1?**

2. **Is Concorde's ground speed always the same when the plane breaks the sound barrier?**

Solution on page 186

What makes a mirror reflect?

Why does a mirror reverse only the left and right sides but not up and down? Alter all, if we speak of a plain mirror, one with a surface that is perfectly smooth and flat, its left and right sides do not differ in any way from its top and bottom surfaces. Why then this persistent preference for changing left to right, while ignoring top and bottom? Can you explain this phenomenon?

While this question is indeed puzzling, we know intuitively that it could not be different. Remember, we are discussing the ordinary mirror, as it is quite possible to construct mirrors that will not reverse left and right, as well as others that will reverse top and bottom.

Q22
Sight and Sound

Why do you have two eyes?

Why are humans endowed with two eyes and two ears? Is the second organ just back-up equipment or a spare, or has it another function?

Q23
Sight and Sound

What makes a sunset's colours?

Sunsets after a blissful summer's day, with the beautiful variety of glorious colours in the fading daylight, are an uplifting experience and the stuff that poems are made of. It is almost blasphemous to try to explain the splendour of this phenomenon in terms of cold physics.

Could you, if you tried?

Solutions on page 187

Why doesn't my kettle make a noise?

There is nothing more cheerful than the sound of a kettle coming to the boil and the prospect of a refreshing drink to come. Have you noticed that the kettle goes strangely quiet just before the boil begins though, and can you explain why?

How do you magnify glass?

Home alone, and gazing out of your window on a sunny afternoon, you suddenly notice a car parked in front of your local bank, with a driver at the wheel. Fumes from the exhaust tell you that the engine is running. Something seems amiss, but you don't want to alert the police. After all, it is only a hunch. That is when you realize that the licence plate is in your line of vision. Unfortunately, you are short-sighted and left your only pair of glasses at the office. However you strain, you cannot see the numbers, and you don't own binoculars. Can you think of a way to enable you to read the licence number from your window?

Solution on page 187

Why does the ringing change?

I was recently putting in some fencing in a field. While knocking in the fence posts, I noticed differences in the sound. In the centre of the field I was a long way from anything, and the sound from the hammering of the posts was dull and flat. I concluded that this was because there was nothing near to reflect the sound. In another part of the field I could hear a distinct echo from a nearby building.

However, in a third part of the field I could hear a distinct ringing tone.

Can you think what might have been the cause of the ringing sound?

Q27
Sight and Sound

Death fall

Frequently in movies someone falls from a high place and we hear a long continuous cry of horror growing fainter as the faller recedes from us downwards.

What is often the common error in this sound effect?

Q28
Sight and Sound

Sound reasoning

Most people know that light, like sound, is a wave form. Yet light and sound behave very differently. Can you explain:

1. **Why it is you can hear around corners, but not see around corners?**
2. **Why you can use sound to cancel out other sounds, but cannot do the same with light?**

Solution on page 188

Q29
Sight and Sound

Why are clear things coloured?

Rayleigh scattering is named after the Nobel prize-winning British physicist **Lord Rayleigh (1842-1919)**. Alongside this phenomenon, which explains every child's questions of why the sky is blue, hediscovered argon, made numerous contributions to humanity's understanding of fluid dynamics and studied transverse surface waves in solids, now known simply as Rayleigh waves. While a member of the House of Lords, he tended only to speak when politics attempted to interfere in matters of science.

Clean air, water, and transparent glass are supposed to be colourless. Yet the sky is blue, a sheet of glass looked through sideways is green, and a mountain lake, pure as it is, appears blue.

Textbooks in physics will talk about "Rayleigh scattering", "electric field", and "electron oscillations", but there is a simpler, if unscientific, explanation.

Do you know what it is?

Solution on page 188

Space and Celestial Bodies
Chapter 2

Q01
Space and Celestial Bodies

Can you square a circle?

Mark and Scott Kelly are twin NASA astronauts who took part in a study of their DNA after Scott had spent nearly a year in space. A flurry of media outlets breathlessly reported that Scott's DNA had changed by 7% in comparison to his brother's, indicating long-term space travel could cause severe mutations. This was swiftly debunked when NASA confirmed there were simply changes noted in relation to 7% of his gene expressions – similar to the effects of mountain climbing or scuba diving.

Prolonged weightlessness can have unfortunate physiological effects on the human body and may prove a great problem on long space flights. It is well known that spinning the spacecraft can produce an effect very similar to gravity, so we have seen science fiction films where the astronauts live in circular living quarters which are rotating to provide apparently normal gravity.

However, as humans we are not used to living in round rooms. What would the effect be if the space explorers were put into a normal rectangular room rotating about a line drawn across the middle of the ceiling?

Solution on page 189

Why aren't
I floating?

An astronaut awakes in an enclosed room where gravity is apparently normal. He realizes that there are three possibilities:

1. He is subjected to gravity.

2. The spacecraft is accelerating.

3. The spacecraft is creating its own "artificial gravity" by spinning.

Is there any experiment that the astronaut can do inside the room to discover which is the true situation?

Solution on page 189

Why do I fly back to Earth?

Most man-made satellites will eventually return to Earth, because their orbit is affected by the Earth's atmosphere which, though extremely thin, still exists very high up. However, the surprising effect of this air-drag is not, as one would expect, to slow the satellite down. On the contrary it will accelerate in its orbit.

How do you explain this phenomenon?

Solution on page 189 **45**

Can you propel a boat with a rope?

Assume that a rope has been tied to the stern of a small boat floating in still water. Is it possible for someone standing in the boat to propel it forward by jerking on the free end of the rope?

Now consider a space capsule drifting in interplanetary space.

Could it be propelled by similar means?

Solution on page 189-190

Q05

**Space and
Celestial Bodies**

How long to travel
through space?

**The nearest star to Earth is Alpha Centauri,
which is about 4.3 light-years away.**

Assuming we can travel at the speed of light, what determines
the amount of time our intrepid voyagers would need to
reach the star?

Solution on page 190 **47**

Q06
Space and Celestial Bodies

Are there rainbows on the Moon?

Rainbows, an enchanting spectacle of nature, pose a number of baffling questions. We know they are caused by drops of water falling through the air, refracting sunlight in such a way that it creates a curve of light exhibiting all colours of the spectrum in their natural order. Now answer the following:

1. Why is it you do not always see a rainbow when it rains while the Sun shines?
2. Can the Moon also produce a rainbow?
3. Even without rain you can, at times, see a rainbow if you look across a lawn early in the morning. Why?

Solution on page 190

Q07
**Space and
Celestial Bodies**

Why does the
Sun go out?

Eclipses of the Sun occur when the Moon moves in front
of the Sun. The Moon takes about four weeks to orbit the
Earth. However, very few people have actually experienced
a total eclipse of the Sun.

**Why is this? Why is there not a solar
eclipse monthly?**

The space station

The International Space Station's first component was launched into orbit in 1998, with the first long-term residents arriving in November 2000. It is now constantly inhabited. The space station can often be seen from Earth – it is the third-brightest object in the night sky – but you have to be quick as it travels at 5 miles per second, circling the planet every 90 minutes.

Weightlessness in a space vehicle is highly inconvenient to any astronaut in many ways. For example, he cannot pour liquid into a cup, neither can he drink from it: controlled movement is possible only by the use of handrails and so on.

It has been suggested that the space stations of the future for use of manned observatories or as staging posts for space exploration might be built in the form of huge wheels with hollow rims, as illustrated. These would be set in rotation so that the outer rim, which acts as the floor, would apply a radial centripetal force to the occupants or any objects inside to keep them moving in a circle.

The equal and opposite reaction to this centripetal force which any person or object exerts on the floor would act as an artificial weight, thus allowing eating, drinking, and working in comparative comfort. The quantum of this weight could be made equal to or less than the normal Earth weight simply by adjustment to the speed of rotation.

1. What would happen to an astronaut's weight if he were to walk round the space station in the direction of its rotation and then turn round and walk in the opposite direction?

2. Now suppose you are in a small windowless room in the same space station after suffering a bout of amnesia. In other words, you do not remember that you are in space. The speed of rotation around the hub produces a simulated gravity of one g. Inside your room everything seems "normal"— gravity seems to be operating on you exactly as it would on Earth. In fact, as far as your senses tell you, you are on Earth.

 In your pocket you have a magnet, a piece of string, some coins, a pencil and a steel paper clip. Suddenly you have some doubt as to where you are. Is there a simple test you could do in your room, using one or more of the objects in your pocket, which would confirm that you are on a spinning space station and not on Earth?

Solution on page 190-1

Pull of the Sun

Surely the Sun's gravity pull on the Moon is much larger than the Earth's. Why then does the Sun not pull the Moon away from its orbit around the Earth?

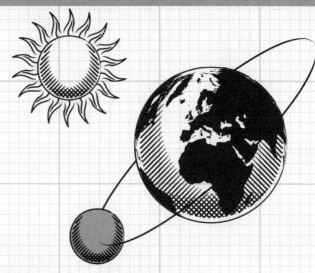

What colour is the Sun?

This might seem incredibly easy to answer, but how many different answers can you think of to the question?

Does the Earth move?

Seismographs are sensitive instruments for detecting earthquakes. They consist of a large mass suspended by springs. When an earthquake happens, the disturbance travels through the Earth, making the instrument vibrate. However, the mass tends not to move because of its inertia. This difference in movement is amplified and written out as a trace.

Can you explain why most places in the world receive two traces for each single earthquake event? Can you also explain why a few places detect only one?

Solution on page 191

Drops and bubbles

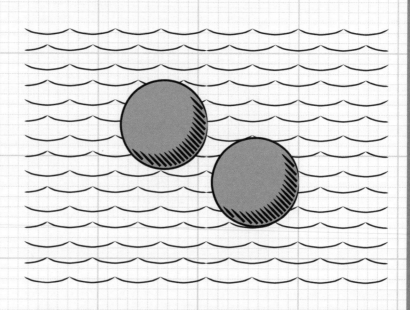

If all space were empty except for two drops of water,
the drops would be attracted to each other, according to
Newton's Law of Gravity.

**Now suppose all space were full of water
except for two bubbles. Would the bubbles
move apart, towards each other, or not
at all?**

Solution on page 191-2

Q13
**Space and
Celestial Bodies**

How does the
Moon cause tides?

The motion of the Earth and Moon causes the tides, and the ceaseless sloshing of the tides is having an effect on the motion of the Earth.

What is happening?

Solution on page 192

Q14
Space and Celestial Bodies

How do you photograph the Moon?

If you want to take a picture of the full Moon and get the largest possible image on film, should you shoot it when it is directly overhead (and therefore at its closest point to your position on Earth), or when it is down hear the horizon? Almost everyone says that the Moon is largest near the horizon. Is this an atmospheric effect or a psychological one?

Does it show up in photographs?

Can you see all of the Moon?

It is common knowledge that we on Earth see only one side of the Moon. In other words, the Moon rotates at a rate around its axis which synchronizes it with the Earth's rotation.

Is this a coincidence?

Q16
Space and Celestial Bodies

Ring around

Suppose we created a ring made of some strong material all around the world supported by small pillars. Furthermore, suppose that, at a given signal, all the pillars are removed at the same time.

Would the ring remain, apparently floating?

Q17
Space and Celestial Bodies

How fast can a space rocket travel?

Cars have a maximum speed. If a driver keeps the accelerator pressed down, the speed increases up to that maximum. While a rocket fires its engine in outer space its speed will increase, no matter how fast it is already going.

Why does a car have a maximum speed, but a rocket does not?

Solutions on page 193

Q18
Space and Celestial Bodies

The spy satellite

After the first Gulf War, the United States wanted to keep a close and constant watch over Iraq. To achieve this objective, the U.S. made use of spy satellites powered by solar cells. These used long-range cameras to transmit important information to their intelligence agencies on Earth.

The ideal solution would have been a satellite hovering over Baghdad 24 hours a day. Instead, the U.S. used a number of them in orbit, changing places as in a relay race.

Why?

Solution on page 193

Q19
**Space and
Celestial Bodies**

How do you relay information in space?

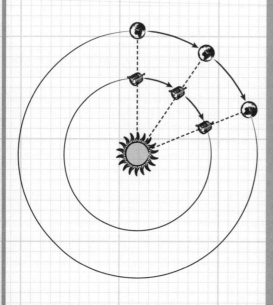

The Solar and Heliospheric Observatory is a spacecraft built as an obeservatory to study the Sun, from its deep core to its solar wind. Launched in 1995, it was originally meant to last just two years, but SOHO is still active in 2019. Its study of "space weather" is of particular importance because of society's increasing reliance on electronic communication; a particularly strong solar flare could damage our communication infrastructure. SOHO has also discovered more than 3,000 comets.

I once watched a news item on television about the launch of the SOHO observatory satellite. There was a computer graphic showing that it was to be placed in an orbit halfway between the Sun and the Earth, with exactly the same period of rotation as Earth. From that position it could observe the relatively nearby Sun and relay information back to the Earth.

What is wrong with this explanation?

Solution on page 193

Shape of the flame

What would be the shape of a candle flame inside a spacecraft where there was normal air but no gravity?

Moonshine

We all know that moonlight is only reflected sunlight. However, when the Moon is starting to wax and the new Moon is only a narrow crescent, we believe that we can also just about see the remaining darkened section of the Moon's face.

Are we simply imagining it because we know it is there?

Q22
Space and Celestial Bodies

Can you steer in space with a shotgun?

You are the sole astronaut on a scientific mission in a spaceship. You rely on solar cell panels to provide energy, but discover that the system is malfunctioning. Repair should not be difficult, so you don your spacesuit and go outside.

As you hammer away, the umbilical cord becomes detached and the spacecraft moves slowly away from you. Not a pleasant situation, particularly as you are not equipped with directional jets, which would enable you to catch up with the ship.

You do have, however, a handgun with a 12-bullet magazine which is able to fire in space. If you were to fire the bullets in one-second intervals in the opposite direction, would you be able to approach the spaceship? If so, would it be at uniform speed?

Suppose the craft is 1,000 metres (3,300 feet) away when you start firing. What precautions, if any, would you have to take?

Solution on page 194

The asteroid

People have speculated that one day in the far future it may be possible to hollow out the interior of a large asteroid or Moon and use it as a permanent space station. Assuming that such a hollowed-out asteroid is a perfect, non-rotating sphere with an outside shell of constant thickness, would an object inside, near the shell, be pulled by the shell's gravity field toward the shell or toward the centre of the asteroid, or would it float permanently at the same location?

Solution on page 194

How high can I jump on the Moon?

The Moon's gravity is about one-sixth of that of Earth. If you weigh 54 kilos (120 pounds) on your bathroom scale, you'll weigh a mere 10 kilos (20 pounds) on the Moon. An object tossed straight up on the Moon will go six times as high as it would on Earth if thrown with the same force.

The first Lunar Olympics will be held in an enclosed arena, warmed and air-filled, so that athletes won't be encumbered by space suits. If a high jumper can clear a 1.8-metre (6-foot) bar on Earth, to what height will he or she be able to jump on the Moon? The answer usually given – 11 metres (36 feet) – is incorrect.

Q25
**Space and
Celestial Bodies**

Does the Earth revolve around the Sun?

You have an academic background and are well read. You have studied the writings of Copernicus (1473–1543) including *De Revolutionibus Orbium Coelestium* (in translation), and of course you know all about Galileo Galilei and his heliocentric view. But you are a sceptic and you wonder whether their theories can be accepted as the ultimate truth. It could just be that Ptolemy and his geocentric blueprint were right. After all, Galileo did recant, admittedly under pressure from the Church, and who knows if the story of his final words on his deathbed, "And yet it moves", is not merely an anecdote.

Now, suppose that money were no object and you had access to the most precise instruments. Could you think of at least three experiments that would prove to your satisfaction that Galileo was right, and it is the Earth that revolves and orbits the Sun?

Solution on page 194

Q26
**Space and
Celestial Bodies**

Why do you get more meteors in the morning?

On any clear night a shooting star can be seen in the sky about every ten minutes, on the average. It is commonly observed, however, that many more meteors are seen in the early-morning hours, between midnight and sunrise, than are seen in the evening between sunset and midnight. An astronomy student tells you that he thinks the reason for this is that artificial lights in homes and offices are usually turned off in the wee morning hours, which makes the whole sky more visible. What do you think of the student's reasoning?

Can you think of a better explanation for reports of morning meteors?

Do Australians observe the same phenomenon, or is the situation reversed "down under"?

Murchison Meteorite

In September 1969 a fireball was seen in the sky near Murchison, Australia. The meteorite split apart, but when all of the fragments were gathered more than 100kg was found. The rock is rich in organic compounds and has been studied in great depth because its composition remains as it was when it was first formed at the birth of our Solar System, roughly 4.6 billion years ago.

Solution on page 195

Q27

**Space and
Celestial Bodies**

Why are there two tides per day?

The ocean tides are caused by the gravitational pull of the Moon and Sun, mostly of the Moon. But why are there two high tides a day? When there's a high tide on the side of the Earth closest to the Moon, there is simultaneously a high tide on the opposite side of the Earth, farthest from the Moon.

How do you explain the second high tide? What makes the water bulge in a direction away from the Moon?

Solution on page 195

Hot and Cold
Chapter 3

How do I make a toy boat?

When, as a boy, I built my own little steamboat to operate in the bathtub, I was thrilled. It was my own design, using odd bits of metal plus a little fatherly help, and looked something like the illustration below.

The principle was simple. The candle heated the water in the boiler and the resulting steam forced the water through the outlets as jets, propelling the boat forward. To my surprise I noticed that, from time to time, the outward jets stopped and instead water was sucked in through the tubes. I did not understand why this should be and, more surprising, why the boat did not then reverse direction and move backwards, as a basic law of physics would suggest (Newton's Third Law of Motion: action = reaction).

Can you explain?

Solution on page 196

Why is the safety lamp safe?

During the 18th century, miners were at extreme risk when exposed to explosive gases. It was in 1815 that Sir Humphrey Davy (1778-1829) invented a safety lamp to be used in coal mines. A fine metal screen in the form of a cylinder covered the open flame of the miner's oil lamp.

Did the screening prevent explosive gases from entering the safety lamp, or is there a different explanation for the lamp's effectiveness?

Why don't I burn my head?

Having washed my hair the other night, I was trying to dry it as quickly as possible. I therefore had my hair drier on its highest setting. I noticed that the air still felt very cold on my head, but burned my ears.

Why?

Solution on page 196

How do you drink on ice?

There are many stories of shipwrecked sailors dying of thirst or being driven insane by drinking seawater. Eskimos have no source of freshwater.

Does polar ice contain salt, and if so how are the Eskimos dealing with the problem?

Climbing the mountain

We were all climbing the local mountain, and as we got higher it seemed to get colder and colder. John said it was due to our greater exposure to the cold winds. Bob said it was something to do with the more rarefied atmosphere. Tom said it was just imagination and as we were nearer the sun it must really be warmer. Bill said we were farther from the centre of the Earth, which was known to be hot.

What do you say?

Solution on page 196

How do you drive on ice?

Whether you are a good or a bad driver makes all the difference in hazardous road conditions. Assume it is freezing and the road is covered with a layer of ice. Can you answer the following questions?

1. If you want to start the car and all you have to help you is one blanket, will you use it as underlay for the front or rear tires?

2. Do you put the car in low or high gear? And will you quickly accelerate or maintain low speed?

3. If you succeed in moving but the car starts to skid, will you steer into the skid or maintain direction?

Solution on page 196

Q07
Hot and Cold

Why doesn't the snow fall evenly?

Have you noticed how much more snow is proportionally deposited on the sides of posts and poles rather than on the sides of buildings?

Why?

Solution on page 197

Q08
Hot and Cold

Does wind affect my mind?

It has long been accepted that weather has a profound effect on the physical well-being and even the mood of most people. One of the best-known natural phenomena is a warm dry wind blowing down from high mountains into valleys. In German-speaking countries this wind is called foehn, though it has many other names with one thing in common: dry, warm and unpleasant. Foehn or its equivalent is held responsible for severe headaches and, in the extreme, for criminal behaviour in some of those affected by it.

How can a warm wind come down from a cold mountain, reach speeds of close to 75 mph, and have such a physiological impact on many people?

Solution on page 197

Why is Death Valley so hot?

Death Valley is a depressed desert region in southeastern California. Most of the valley is below sea level and it has the distinction of being the hottest place in the world. In 1913 a temperature of 134°F was recorded – the highest ever at the time.

Elementary physics has taught us that hot air rises and cold air sinks. Would you therefore not have expected Death Valley to be a cool or moderately warm place, particularly as it is almost entirely enclosed by mountain ranges?

Solution on page 197

How do you cool off?

Your air conditioner breaks down on the hottest day of the summer, leaving you sweltering in your one-room apartment. In order to get some relief, you leave your refrigerator door open.

Will you cool the room that way?

Why does my tap steam?

Q11
Hot and Cold

I could never understand why, whether at home or in a hotel, after turning on the hot water the flow of the stream first decreases, only to return to normal after about 30 seconds.

Do you know what causes this?

Q12
Hot and Cold

Thermal expansion

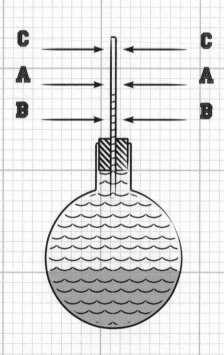

Thermal expansion is the term used to explain the phenomenon involving the increase in volume of a material as its temperature rises. This occurs because the higher temperature causes a higher molecular kinetic energy, which in turn cause a greater average separation between the molecules. The effect is especially problematic in engineering, where it can result in everything from cracked pipes to train derailments.

We know that most materials expand when heated. This property can be demonstrated easily in the interesting experiment below:

A flask, filled with water, is fitted with a rubber stopper, through which a tube is inserted into the flask and topped up with water until the water level rises a short distance up the tube to A-A.

The flask is then plunged into a container of hot water, and now the unexpected happens. The water level in the tube first falls to B–B before it rises steadily to C–C.

Can you explain why?

Solution on page 197-98

Why is the floor cold?

The other night I could not sleep and was feeling hungry. Deciding to raid the refrigerator, I stepped, barefoot, from the carpeted hall into the tiled kitchen and found that the floor was extremely cold. The door between the hall and the kitchen had been open all the time.

So why was the kitchen floor so much colder than the hall floor?

Solution on page 198

Q14
Hot and Cold

Breath on
your hand

If you breathe gently on the back of your hand it feels warm.
If you purse your lips and blow hard on the back of your
hand it feels cold. Surely the temperature of one's breath is
the same on both occasions, so why does it feel different?

Solution on page 198

Blowing out the candle

Take a lighted candle and a sheet of paper, which you roll into a hollow cone, as shown. If you try to blow the candle out by blowing through the small end, you will fail.

How can you be more successful?

Solution on page 198

How do I boil water in my hand?

One of the earliest lessons in elementary physics taught us that water would boil at lower temperatures at reduced pressures and at a higher temperature if the pressure was increased. An experiment to prove this is still vivid in my memory.

An enclosed glass container, as illustrated, containing water in a partial vacuum will boil if you grip one end. My wife believes that this experiment was just a trick by our physics teacher, as in her experience, when preparing the spaghetti the water boils much faster if she puts the lid on the pan, thus increasing the pressure. Is the school experiment just a sleight of hand or is my wife mistaken?

Solution on page 198

Q17
Hot and Cold

Insulating metal pipes

Here is a problem that might require a little more than elementary household physics.

Exposed hot-water and steam pipes are often insulated to reduce heat loss. What is surprising is that the "insulator" is often a better conductor of heat than air; for instance, cork is twice as good as air at conduction. Yet one would have thought that using such a material would be counter-productive. Why do we insulate pipes?

If the thermal conductivity of the insulation is so unimportant, why do we not just use thicker pipes? That is to say, pipes with the same internal diameter but greater external diameter.

Your biggest fan

Q18
Hot and Cold

Why is it an advantage to have a constantly running electric fan in a room heated by a wood or kerosene stove, but not in one heated by a big fireplace?

Solutions on page 198-99

How do you walk on fire?

Fakirs, at home in India, depend on begging for their sustenance. They are generally regarded as holy men and are said to be endowed with miraculous powers, such as walking on fire.

I have often wondered whether these stories are true, and if they are, whether these feats can only be performed under hypnosis or in a state of trance. A similar phenomenon can be observed by performers who lick red-hot metal bars.

Can ordinary physics explain these demonstrations?

Iron doughnut

A piece of solid iron in the shape of a doughnut is heated over a fire. As the iron expands, does the hole in the middle become larger or smaller, or does it remain the same size?

Solution on page 199

How do you open a decanter?

Individuals using perfume bottles with glass stoppers are often confronted with stoppers that cannot be easily removed without using force that could result in breakage. Placing it under a hot water tap won't work. Yet there is an easy method that works without fail.

How would you solve this problem?

Q22
Hot and Cold

Archimedes (c.287-c.212BC) of Syracuse was one of the leading scientists and mathematicians of the ancient world. His contributions were many and include a number of geometric proofs, a device for raising water that transformed irrigation and, of course, the word "Eureka"!

Can you start a fire with a mirror?

According to a famous legend, when Roman ships attacked Syracuse in 214 BCE, Archimedes saved the day with mirrors. He positioned soldiers on the shore, each with a large mirror. At a signal, they burned the Roman fleet by reflecting sunlight into the ships. Fiction aside, is there any practical validity to Archimedes' tale? Whether or not he really tried to defend Syracuse with mirrors, we probably will never know, but could his trick have worked?

Solution on page 199

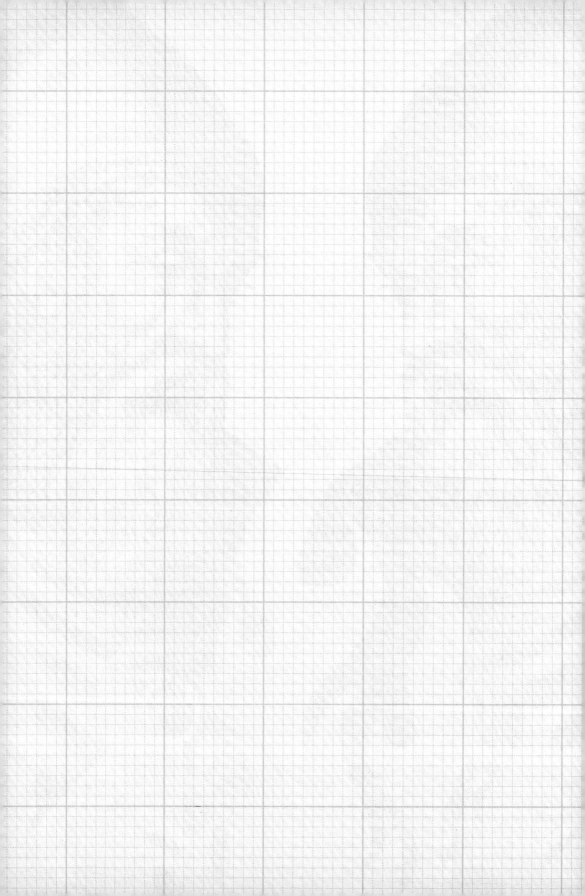

Weighty
Chapter 4

Q01
Weighty

How can I balance my scales?

My kitchen scales are balanced when there is nothing on them, but I know that there is something wrong because I get different results depending on which side I put the object to be weighed. If an object balances with 100 grams when on one side, but with 144 grams on the other side, is it possible to determine the real weight?

Quito

Q02
Weighty

Suspend your disbelief now for this most intriguing puzzle. Approximately how much would a man living in Quito, Ecuador, weigh if he were 44,000 miles tall? Would it be more or less than a thousand tons? (This needs a great deal of thought, not just an inspired guess.)

Solution on page 200

How can I make a spray gun?

As children, it was fun to annoy grown-ups by spraying them with water, using a gadget as illustrated.

By blowing through the horizontal tube, you are forcing water up the conduit tube into a fine spray. The same principle is used in aerosols and paint sprayers.

Why does water rise in the tube, against gravity?

Q04
Weighty

Why don't submarines rest?

The first submarine to work successfully was built for James I by Dutch engineer **Cornelius van Drebbel (1572-1633)**. It was tested on the River Thames around 1620, and it was likely to have been an enclosed rowboat manned by oarsmen. It worked, diving to 15 feet, but unfortunately none of the original plans or designs survive so it is hard to know exactly how this was achieved.

The captain of a submarine tries at all costs to avoid letting his sub come to rest on a clay or sandy ocean bottom. He knows that if this happens, it can be fatal.

Why?

Solution on page 200

Q05
Weighty

How do you steer a boat?

Can you steer a boat on a lake if it is entirely becalmed and you have no oars? The answer must be no!

Suppose the circumstances are the same except that your boat is drifting in a fast-flowing river.

Can you then steer with the rudder?

Solution on page 200-201

Q06
Weighty

Balloon in a bath

An air-filled balloon is held underwater by a weight so that it is just on the verge of sinking; the top surface of the balloon just touches the waterline. If you push the balloon beneath the surface of the water (as shown), what will it do—bob back to the surface, remain at the level to which it is pushed, or sink to the bottom?

Q07
Weighty

How do you walk on a tightrope?

No circus act is complete without a tightrope walker. How does he keep his balance and why does he use a long bar?

Does a bootstrap elevator work?

Q08
Weighty

Study the drawing carefully.

Can the man lift both himself and the block off the ground?

Solutions on page 201

Q09
Weighty

Gravity train

A German professor described a train that would travel from city to city, powered only by gravity: "Each railway is in a long tunnel, perfectly straight [that is, not following the Earth's curvature]; so of course the middle of it is nearer the centre of the globe than the two ends; so every train runs halfway downhill, and that gives it force enough to run the other half uphill."

Ignoring friction and air resistance, as before, would such a gravity train work? And if so, about how long would it take to get from, say, Paris to Madrid? From Los Angeles to Moscow?

Solution on page 201

Hourglass puzzle

You can sometimes find an unusual toy in the shops: it is a glass cylinder full of water, with a sand-filled hourglass floating at the top.

When the cylinder is turned upside down, as in the right-hand drawing, something rather strange happens. The hourglass stays at the bottom of the cylinder until a certain amount of sand has flowed from its upper compartment into its lower compartment, then it rises slowly to the top.

Can you suggest a simple explanation for this phenomenon?

Solution on page 201-2

The half-hidden balance 1

The horizontal line is a weightless rod, balanced on a fulcrum at F, and of unknown length to the right of the fulcrum. The visible part is 1 foot long, and supports a 1-pound weight. Besides being weightless, the rod is able to support any weight, and, since it is in balance, there must be a weight to the right. The further the second weight is to the right, the less it can be.

What are the lower and upper limits of the total possible force downwards at F?

Q12
Weighty

Hole through the Earth

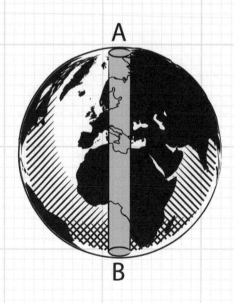

A

B

The internal structure of the Earth is layered in spherical shells, from the solid crust, to a viscuous mantle, a liquid outer core and finally a solid inner core. The inner core was discovered in 1936 by the Danish seismologist **Inge Lehmann (1888-1993)**. She analysed the seismic waves generated by earthquakes in New Zealand, realising the only explanation for them was reflection from a solid material in the core. The temperature at the inner core is estimated to be 5,400°C.

Assume that a hole is drilled from one point on the globe through the centre of the Earth to the Antipodes, as illustrated, and a steel ball is dropped into the hole at point A.

Ignoring any external influences such as air resistance, friction and conditions of the Earth's core, answer the following questions:

1. As the ball is travelling from A to the Earth's centre, does its velocity increase, decrease, or stay the same?
2. Will the ball weigh less or more when it reaches the centre of the Earth?
3. Will the ball's mass change during its journey?
4. At what point will the ball be subject to zero gravity?
5. If the ball fell through such a cylindrical hole through the centre of the moon, would the one-way journey take more or less time than it would on Earth?

Solution on page 202

Rope and pulley

The diagram shows a rope passed over a frictionless pulley. To one end a weight is suspended, which exactly balances a man at the other end. What happens to the weight if the man attempts to climb the rope? Assume the rope to be weightless and the wheel to be frictionless.

The tank

A closed glass tank is completely filled with water. A cork sphere is fixed to the bottom by a thread to prevent it from rising further. A steel sphere hangs from a thread fixed to the top of the tank. The lengths of the threads are such that the steel sphere is hanging just above the cork sphere.

What happens if the tank is suddenly moved to the right?

Solution on page 203

The bottle and the coin

A well-known parlour trick is shown below:

A small coin is put on a card (e.g., a visiting or business card) and placed over the mouth of a bottle. When the card is flicked away with the finger, the coin drops into the bottle.

Can you explain why this should be so?

Solution on page 203

Fishy business

I have been watching my goldfish swimming around in its bowl, and have noticed that occasionally it swims to the surface to take a gulp of air. I know the effect would be very small, but wondered if this would cause the bowl to increase or decrease in weight, or stay the same.

While thinking about it, I wondered if I could simulate this effect by using an underwater syringe to draw in air from the surface like a kind of mechanical fish.

What do you think?

What does an hourglass weigh?

You have an hourglass which has run its course on the table in front of you. What happens if you turn it upside down? Will the weight of the hourglass be reduced while some of the sand is in free fall?

The half-hidden balance 2

In the diagram, there is a cylindrical iron bar, one square centimetre in cross-section, suspended vertically over a beaker. The beaker is two square centimetres in cross-section and partly filled with water. The bar just touches the surface of the water. The beaker is standing on the right-hand pan of a balance scale. The left-hand pan contains an empty beaker and sufficient gram weights to balance the scale. One cubic centimetre of water is added to the right-hand beaker.

How much water must be added to the left-hand beaker to bring the scale back to balance?

Solution on page 203

How do you skim stones?

The ability to skip flat stones across the water is a question of skill. It is difficult to measure the path of a stone across water, but if you skip a stone on the wet sand at the water's edge, it will leave marks tracing its path. The flight is surprisingly complex. Long hops of several feet alternate with short hops of just a few inches, and zigs to the left alternate with zags to the right. A right-handed throw, with the proper grip, spins clockwise and strikes the sand first with its trailing edge.

Will the first hop be short or long? To the left or the right?

Q20
Weighty

Pendulum clocks

Christian Huygens (1629-1695) was a Dutch physicist and inventor, who is widely regarded as one of the greatest scientists of all time. Among his many inventions and discoveries was the pendulum clock in 1656, a remarkable breakthrough in timekeeping that changed the world.

If, in a vacuum on Earth, you hang two objects of the same weight by two strings of the same length they will swing in time with one another. If you increase the length of one string its weight will swing more slowly.

What will happen if instead you increase one weight?

Solution on page 204

Q21
Weighty

How do you burst a barrel?

There used to be a trick popular in Victorian times. A beer barrel was completely filled with water. A long thin tube had been attached to the top of the barrel, which was empty at the start of the trick. By pouring a very small amount of water into this tube the barrel could be made to burst, demonstrating that small causes can have large dramatic effects.

Suppose one jug of water was needed for the trick. If a narrower tube with half the cross-sectional area had been used, how much water would have been needed?

Q22
Weighty

The wheelbarrow

Which is easier: pushing a wheelbarrow or pulling it?

Q23
Weighty

A question of stability

Scientists recognize that objects can have three types of stability:

1. stable – tilt the object a bit and it returns to its original position, e.g., a cylinder sitting on its base. (Note that a cylinder has two stable positions because it has two bases.)
2. unstable – disturbing the object just a little makes the object move to a completely new position, e.g., a cylinder balanced on an edge. (Note that a cylinder has two unstable positions because it has two edges.)
3. neutral – rotate the object and it stays in its new position, e.g., a cylinder on its side. (Note that a cylinder has one neutral position because it has one side.)

We could call a cylinder a 2:2:1 object, because that is the ratio of its stable : unstable : neutral equilibria. Can you find examples of the following objects:

a) 1:2:1 b) 0:0:1 c) 0:2:1 d) 1:1:0

Can you prove that 0:x:0 is impossible?

Solution on page 204-5

What's in the box?

The box illustrated is resting on a table, projecting slightly more than half its width over the edge. It can remain in this position owing to the weight of something inside it. When the table is bumped vertically, the box falls. What does the box contain? Its contents can be found in most households.

Imagine a similar box, but projecting less than halfway over the edge of the table. In this case it can be made to fall without bumping, touching, or applying any pressure to it or to the table, directly or indirectly.

How? What is in this box?

How do you balance a ruler?

Hold a ruler horizontally, balanced across your first two fingers. Then try to move your fingers together so that both slide. It does not work. The ruler first slides over one finger and then over the other, but never both together.

Do you know why?

How can I get off the table?

I am at rest on a perfectly smooth, frictionless table, so that the only force acting on me is the upward pressure of the table due to my weight.

I can only shift my centre of gravity up and down, and therefore I cannot crawl. The table is also so large that I cannot seize an edge. How do I get myself down off the table?

Q27
Weighty

The perforated water can

Drill a number of small holes in a can and fill it with water.

1. Fit an airtight lid and the water stops leaking from the holes. Why?
2. Drill a hole in the lid and the water now flows. Why?
3. Remove the lid, run your finger through the leaking streams, and, if the holes are not too far apart, they will merge as shown, forming a single stream even after you have removed your finger. Why?

Solution on page 205

Q28
Weighty

The bucket

Consider a bucket of water with two holes of equal area through which water is discharged. The water can flow out through hole (B), at the bottom, or through the downspout, which begins at the top (T) and has its opening the same distance below the water level as the centre of hole (B).

Ignoring any friction effects, out of which opening will the water flow faster, and why?

Solution on page 205

121

Q29
Weighty

The mystery of weight

The theory of gravity is one of science's best-known theories, first hypothesised by English physicist **Isaac Newton (1642-1727)**. Gravity on Earth causes an acceleration of approximately 9.8m/s², although the effects of gravity are not uniform around the globe; mountains and valleys, as well as different composition of rocks and minerals, cause small variations in the strength of gravity.

Many volumes could be written on the interconnected phenomena of weight, weightlessness and gravity. Letting our imaginations run wild, suppose we were to stand on a spring weighing machine at the Equator and that, by some means, the Earth's rotation could be speeded up, carrying its atmosphere along with it.

1. Would the weighing machine indicate a larger or smaller weight, or would there be no change?
2. If we speeded up the Earth's rotation further, would we reach a critical point, and if so, what would happen?

Solution on page 205-6

How do you build a bridge of bricks?

A farmer wants to build a bridge made of bricks, over a small river. His son, a student architect, suggests two designs as shown:

1. Which design can carry heavier traffic?

2. Which will need more building material?

Solution on page 206

Centre of gravity

Suppose you are a manufacturer of cardboard boxes. Business is slow, and you are therefore more than happy to receive an inquiry for 50,000 boxes. The specification is somewhat unusual in as much as the shape of the box is to be irregular, as illustrated.

Furthermore, the customer, for his own reasons, wants you to mark the projection of the centre of gravity of the box on the lid with a red dot. In other words, the empty box would balance on that point. You are asked to submit a sample with your quotation. How will you quickly find the centre of gravity? (Note: this is not a geometry problem.)

Solution on page 206

Q32
Weighty

How do you make a cork float centrally?

Take a glass that is partially filled with water and drop a cork into it. The cork won't float in the centre, but keeps drifting to one side and attaching itself to the side of the glass.

Without swirling the water, how can you make the cork float in the centre?

Q33
Weighty

The Archimedes puzzle

Legend has it that Archimedes was asked to verify whether a crown was made of pure gold, as was alleged, or whether it had been adulterated with silver.

Suppose you know the specific weight of gold, how would you go about the problem, short of melting down the crown? Remember that gold is heavier than silver per unit volume (specific weight).

The brick and the dinghy

An inflatable dinghy is floating in a swimming pool. Which will raise the water level higher: throwing a brick of metal into the dinghy or throwing it into the water?

Solution on page 206

Q35
Weighty

Why doesn't the atmosphere crush me?

Léon Philippe Teisserenc de Bort (1855-1913) was a French meteorologist who is credited as co-discoverer of the stratosphere (Richard Assmann announced his dicovery during the same time period in 1902). Teisserenc de Bort pioneered the use of unmanned instrumented balloons to measure the atmosphere and used these to discover the height at which the air's temperature ceases to fall; an altitude of around 7 miles.

One of the properties of the atmosphere is air pressure, which shapes the physical world in which we live. Because we do not feel it, we don't give it a second thought, and we are blissfully unaware of the profound effect this phenomenon has on our existence. We are not even conscious of the load exerted on us by atmospheric pressure, which for the average-sized man can exceed 12 tons.

Why does this enormous load not crush us?

Solution on page 206

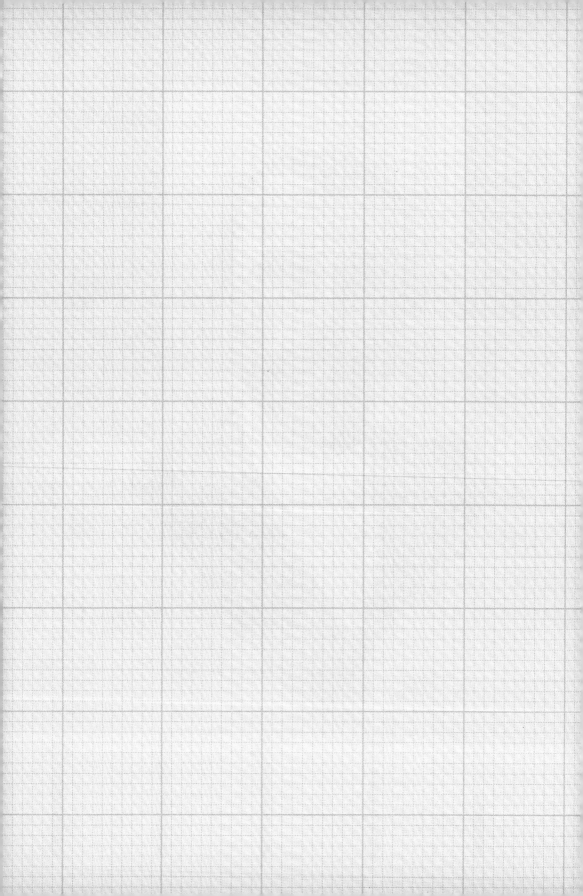

Up in the Air
Chapter 5

Why does a golf ball have dimples?

In the early days of golf the balls were smooth. Dimples were introduced later, after manufacturers claimed the dimpled variety travelled farther.

Were they right, and if so why?

Why don't the clouds move?

I was on a hiking holiday, when I noticed a peculiar phenomenon. At first, I could not work out what it was that "did not feel right". I was walking across a large plain. To the west was a range of mountains that stretched north to south as far as I could see. There was a strong wind blowing from the direction of the mountains.

I suddenly realized what seemed odd. It was the clouds. They were an unusual shape, forming long straight bands parallel to the mountains. But the really peculiar thing was that, despite the strong wind, they did not appear to be moving at all. Can you explain?

Solutions on page 207

How do birds fly?

Did you ever ask yourself what it is that enables birds to fly? After all, they are heavier than air and therefore do not float. To fly they would need lift and forward propulsion.

How do birds manage to achieve this?

Q04
Up in the Air

Why don't planes fly lower?

According to a study undertaken in 2017, at any given moment there are an average of **9,728 commercial airplanes** in the sky around the globe. That equates to well over one million people in the air at the same time. The heaviest travel day during the year-long period studied was 5 August 2016, which saw a peak of 12,856 planes flying at once. Conversely, the lightest day for air traffic was 1 January 2017, which saw a low of just 3,3,54 commercial planes in the sky.

We know that as the radius of a circle increases, so does its circumference. This must mean that the higher an airplane flies, the greater the distance it has to fly to its destination.

On a flight between London and New York, an airliner has to fly an extra two miles because of its high altitude.

Forgetting air-space congestion as a possible argument, why do airliners not fly by a shorter, lower route?

Solution on page 207

Why do boomerangs come back?

The boomerang is a missile used mainly by Australian aborigines. Originally used as a weapon, tossing the boomerang has recently become a sport. It is generally held vertically in the right hand, although some are designed for left-hand use. The fascinating aspect of this weapon is its ability to return to the hands of the thrower.

Can you explain this phenomenon?

Solution on page 208

Q06
Up in the Air

Does the balloon move?

A child sits in the back seat of the family car, holding a helium balloon on a string. All the windows are closed. As the car accelerates forward, does the balloon tilt forward, tilt backwards, or remain in the same place?

Can you fly a jet-powered helicopter?

Q07
Up in the Air

In airplanes, the lift is supplied by the wings while the thrust is supplied by the engines. Jet engines can make the plane move very fast indeed. In helicopters, the lift is supplied by the rotors, which also provide the thrust by tilting the nose down. Could we get helicopters to go as fast as a plane if they were propelled by jet engines, while the rotors were used solely to provide lift?

The flag and the weather vane

Assume that the wind blows with uniform force from the same direction. A weather vane will remain stationary while a flag, even if perfectly smooth and fully spread by hand to start with, will flap once released.

Why?

Should I have a fear of flying?

My wife hates everything about flying. Taking off, landing, the noise of lowering the undercarriage and the change of pitch on acceleration or deceleration of flying speed.

While I managed to calm her down, explaining that all this is standard procedure, I am at a loss when we fly through a thunderstorm with lightning all around us. In fact I am a little concerned myself, although I do not tell her so.

Why is there no catastrophic damage if lightning strikes a plane?

Q10
Up in the Air

Two cannonballs

Two identical cannons are aimed directly at each other, as in the illustration. The only difference is that the cannon on the left is at the top of a cliff and shooting downwards, the one on the right is aiming up.

Neglecting air resistance, what will happen if the perfectly aligned charges are fired simultaneously and at the same speed?

Solution on page 209

Q11
Up in the Air

A bolt from the blue

Benjamin Franklin (1706-1790) was an American scientist and founding father who was instrumental in the early exploration of electricity. He developed a test to discover whether lightning was electricity. This did not, as is popularly depicted, involve flying a kite in a storm. Instead he used the kite to collect electrical charge, which he then stored in a Leyden jar.

A report in the newspaper the other day told the story of a farmer who had had a lucky escape. He was in a field when a nearby tree was struck by lightning. At the time he happened to be standing next to a cow that promptly fell down dead.

Had the cow been suffering from a weak heart, or is there another more scientific reason?

140

Solution on page 209

Q12
Up in the Air

Can you blow up a balloon?

You have two identical balloons, except that Balloon A has been blown up to 10cm (4 inches) diameter and Balloon B to 20cm (8 inches).

You want to blow ten times into each balloon. Assuming neither bursts, will it be easier to succeed with one balloon or the other, or will the necessary force be the same?

Q13
Up in the Air

How do you float a balloon?

A hydrogen balloon will rise into the air. Suppose that we wanted a balloon that neither rose nor sank, but hovered at a constant height. We will do this by attaching a length of string so its weight is just enough to balance the upward force. You could do this by starting with a piece of string which was too long and successively snipping bits off until it just balanced. The trouble is that if you cut off too much, you will have to start all over again. Is there a simple way to achieve the desired effect?

Solutions on page 209-10

Q14
Up in the Air

Dropping a bombshell

You may have seen films of planes dropping bombs to which small parachutes have been attached. Why is this done? Surely it is not to ensure that the bombs drop gently?

Q15
Up in the Air

The stalling plane

If a plane stalls, the pilot will be unable to pull out of the dive immediately, but is likely to succeed after having nose-dived for some time.

Why?

Solutions on page 210

Q16
Up in the Air

Free fall

A cork is suspended in the middle of a bucket of water held down by a fixture to prevent it rising to the top, as shown in the illustration.

The bucket is then dropped, with the fixture remaining on the roof, so that the cork is free to move in the water. Where will the cork be just before the bucket hits the ground?

Solution on page 210

How do you make a truck fly?

There is a well-known, probably apocryphal story:
A truck full of live poultry is stopped before a bridge on a country road. Its driver, beating the side of the truck with a stick, is asked what he is doing and he explains that his load is too heavy for the bridge so he is making the birds fly to lighten his load before proceeding.

This suggests the following puzzle. A cage with a bird in it, perched on a swing, weighs four pounds. Is the weight of the cage less if the bird is flying about the cage instead of sitting on the swing? Ignoring the fact that if left in an airtight box for long the bird would die, would the answer be different if an airtight box were substituted for the cage?

Q18
Up in the Air

How does a helicopter fly?

Helicopters have been theorised for many centuries, with the most famous early design originating with Leonardo da Vinci in the 1480s. However, it was not until Russian pioneer **Igor Sikorsky (1889-1972)** created his experimental helicopter in 1939 that successful hovering was achieved. He achieved free flight a year later, and his helicopters were used by the Allied forces in World War Two.

We are all more or less familiar with this heavier-than-air aircraft, which has one or more power-driven horizontal propellers that enable the craft to take off and land vertically, move in any direction, or hover stationary in the air. However, you may not have noticed the small rotor at the helicopter's tail.

Is it important? What do you suppose is its purpose?

Solution on page 211

In the Lab
Chapter 6

How can I tell if an egg is raw?

Q01
In the Lab

You have an egg before you. How can you find out whether it is hard-boiled or raw – without, of course, opening it?

Q02
In the Lab

How do you make salad dressing?

Jill and Joe went out for a picnic. "I hope you remembered to bring the oil and vinegar for the salad," said Joe.

"I certainly did," replied Jill, "and to save myself having to carry two bottles I put both the oil and vinegar in the same bottle."

"That wasn't very clever," said Joe, "because, as you very well know, I like a lot of oil and very little vinegar, but you like a lot of vinegar and hardly any oil."

Jill sighed, and then proceeded to pour, from the single bottle, exactly the right proportions of oil and vinegar that each of them wanted. How did she do it?

How do I bisect an object?

Suppose we have a number of flat objects, each with a different shape: square, circle, and so on. If we select any of those objects we could draw a bisector across it. I will define a bisector as a straight line that divides the object into two halves. Of course for any object we could draw an infinite number of bisectors. Would all the bisectors necessarily go through the same point for any specific object?

Q04
In the Lab

How do you tell acid from alkali?

Litmus was used for the first time by **Arnaldus de Villa Nova (c.1240-1311),** an Aragonese physician and alchemist. It occurs naturally in various lichen, and is used to test whether a solution is acidic or basic. This works because of the molecule 7-hydroxyphenoxazone, which contains hydrogen ions that disassociate as the pH increases, causing a change in colour from red to blue.

Litmus paper is widely used for laboratory purposes. It consists of an absorbent paper impregnated with a dye, and is the oldest form of acid-base indicator. Dipped into an acid solution it will turn red, and in a base solution it will turn blue.

If no litmus paper is available, do you know a natural product which will serve the same purpose?

Solution on page 212

Q05
In the Lab

Can you blow up a room?

You are in a room filled with 100 percent methane gas. What would happen if you struck a match?

Q06
In the Lab

Whisky and soda?

You have a litre of soda water and a litre of whisky. Take one tablespoonful of soda, transfer it to the whisky, and mix thoroughly. Then take a tablespoonful of this mixture and pour it back into the soda. Is the amount of soda in the whisky greater than, less than, or equal to the amount of whisky in the soda?

Solutions on page 212-13

Q07
In the Lab

How do you get an egg into a bottle?

To put big objects into bottles has become something of an art. It seems as impossible as passing a camel through the eye of a needle, but these objects include such items as pears, arrows, and ships. Various techniques are used that might differ from object to object.

Our puzzle deals with a classic of this type, namely a hard-boiled egg into a milk bottle.

The method is pretty well known. First you drop a piece of paper and a lighted match into the bottle, then place a peeled egg upright over the neck. The oxygen consumed by the fire creates a partial vacuum and the atmospheric pressure will push the egg into the bottle. After you wash the ashes and dead match out, you can take a bow.

But, can you do the same with an unpeeled egg?

Solution on page 213

Perpetual motion I

To avoid misconception, it is important to understand the essence of perpetual motion and to realize that such a contraption would violate the laws of nature. Nonetheless, many such machines have been proposed throughout history. Illustrated is one ascribed to the Bishop of Chester (1674).

A magnet is fixed atop a column and will draw a steel ball up the ramp until it reaches the cut-out. It would then drop to the base of the ramp, only to be attracted once again by the magnet to repeat the performance.

Where is the fallacy?

Solution on page 213

Q09
In the Lab

Tables

You live in a hilly section of the suburbs and want to buy a new set of table and chairs for the garden. Would you be better off buying items of furniture that have four legs or three legs?

Make a rainbow

Q10
In the Lab

One of the most beautiful sights in our physical world is that of soap bubbles showing all colours of the rainbow. Is the air pressure inside the bubble larger or smaller than, or equal to, the atmospheric pressure outside the bubble?

Logic will provide the answer.

Solutions on page 213

Q11
In the Lab

The drinking bird

Many toys and promotional gadgets are based on physical phenomena, often not readily understood. The drinking bird is one of the most fascinating examples. It consists of a glass bird, standing in front of a container filled with water.

The bird appears to be thirsty at regular intervals and so it rocks forward and dips its head into the water. Having quenched its thirst, the bird rights itself, only to repeat the performance after a little while. All you have to do to get the operation going is to wet the felt that is covering the bird's head and beak, after which the bird continues to bob up and down without further assistance.

What makes it do this?

Solution on page 214

The leaning tower of dominoes

Stack dominoes on top of each other so that each domino projects as far as possible over the domino below without falling. If you stack two dominoes, the top one can overhang by as much as 50 percent of its own length. If you stack three dominoes, the greatest offset is obtained when the centre of gravity of the top domino (A) is directly above the end of the domino below (B) and the combined centre of gravity of these two dominoes is directly above the edge of the third domino (C). The top domino now overhangs the bottom one by 75 percent of its length.

By continuing this pattern, is it possible to build a tower in which the top domino projects more than a full domino length beyond the bottom domino? If so, approximately how many dominoes do you think will be required to achieve this?

Solution on page 214

Tumbling polyhedrons

Each face of a polyhedron can serve as a base when the solid object is placed on a horizontal plane. With a regular polyhedron (all faces alike), the centre of gravity is always directly above the centre of a face; therefore it is stable on any face, as illustrated.

Irregular polyhedrons that are unstable on certain faces are easily constructed; that is, when placed on a plane with an unstable face at the bottom, the centre of gravity is not directly above that base, and the polyhedron topples over.

Is it possible to make a model of an irregular polyhedron that is unstable on every face?

Soap bubbles

Imagine two soap bubbles of differing size blown on the ends of a T-shaped tube. The blowing inlet is then closed, leaving an air passage connecting the two bubbles through the tube. What will happen? Will the smaller bubble expand at the expense of the larger until they are both the same size? Or will something else happen? Or will nothing happen?

Solution on page 214-5

The spinning top

The most famous spinning toy of all is the top, and the most perplexing version of it is the Tippe Top (*left*).

Its behaviour is so paradoxical that whole treatises about it have appeared in *Physica*, *The American Journal of Physics*, and other scientific periodicals. Physicists are still not in agreement as to how it works.

The Tippe Top has a spherical bottom and a stem on its top. The surprising thing is that when you spin it on its spherical surface, it stays on this side for only a few seconds, then turns upside down and spins on its stem. This raises the top's centre of gravity, which would seem to violate the principle of the conservation of energy.

You can demonstrate this paradox to yourself with any Tippe Top or with any typical ring that has a smooth stone. Spin the ring on its stone, and in a few seconds the ring will invert itself and spin with the stone facing upwards. A hard-boiled egg that is spun flat will rise to spin on one end. (Spinning an egg is a good way to tell whether it is hard-boiled; a fresh egg won't spin because the liquid inside sloshes around too much. See page 150.)

Do you have an explanation?

The mercury experiment

Do not try this experiment. It is included in this book only for discussion. If you could fill a glass test tube, open at one end, with mercury, then close the open end with your finger and then invert the tube into a vessel filled with mercury, as shown in the illustration below:

1. **What happens when your fing is removed?**

2. **What practical use is made this experiment?**

Tongue-in-Cheek Physics

Chapter 7

Q01
**Tongue-in-Cheek
Physics**

Perpetual motion 2

A weight is suspended over a pulley and attached to the ground by a helical spring. As the world turns under the Moon, its gravity makes the weight move up, turning the pulley. The pulley is connected to a generator, creating electricity.

Would the set-up work? Is it a perpetual motion machine? Where does the energy come from?

Solution on page 216

The light
and the shadow

A man is walking along a level road lit only by a solitary streetlight. He is moving at a constant speed and in a straight line as he passes the lamppost and leaves it behind, which makes his shadow lengthen. Does the top of the man's shadow move faster, slower, or at the same rate as the man? Prove your answer.

Solution on page 216 **169**

Q03

Tongue-in-Cheek Physics

Feathers and gold 1

Which weighs more, a pound of feathers or a pound of gold?

Q04

Tongue-in-Cheek Physics

How heavy can you get?

Let us suppose that there is no limit as to how tall a person could grow. Now imagine someone growing rapidly, becoming progressively taller and therefore heavier.

The question the reader is asked to consider is this: Would the person's weight increase forever?

Solutions on page 216-17

Q05
Tongue-in-Cheek Physics

Can you float a battleship in a bathtub?

Specifically, the battleship on the left below, which, floating on the high seas, has a mass of 30,000 tons. Imagine slowly lowering the ship into a huge bathtub that is shaped like the ship but a little larger, containing a small volume of water. As the ship is lowered, the water is forced up the sides of the tub and over the rim until there is only a thin envelope of water between the tub and the ship's hull. Is it possible to float a 30,000-ton battleship in just a few hundred gallons of water? Keep in mind that a floating object displaces a volume of water equal to its weight and that 30,000 tons of water have a volume of several million gallons.

Solution on page 217

Q06

Tongue-in-Cheek Physics

The spying game

In war as in industrial espionage, the main problem is to avoid detection when transmitting messages. Microdots and invisible ink are old hat, and new methods are devised all the time. Paul Goodwin was employed on a top-secret project concerning isotopes and his problem was to transmit a message to his principals abroad. Coded messages were too dangerous and therefore unthinkable. After much consideration he devised what he considered to be a foolproof method.

There are 26 letters in the alphabet. As any message would also contain punctuation marks and blanks (between words), we are dealing with a total of about 40 symbols. Paul assigned a two-digit number to each, starting with 01 for "A", 02 for "B", and so on until he reached 38 for a full stop and 39 for an interval between words. (ISOTOPE MASS would, for instance, read: 09 19 15 20 15 16 05 39 13 01 19 19.) With the help of his powerful computer, Paul Goodwin translated the message into one long number. The reader is now asked to permit some intellectual licence. By putting a decimal point before the number, we can consider the whole code to be expressed as the length in metres of a bar made of precious metal (to avoid any chemical reaction with the atmosphere). This bar would have a length of more than 9 but less than 10 cm.

Let us assume that Paul has the means to cut the bar to the required length, which would require fantastic precision.

Let us also assume that the recipient of the bar, having equally precise measuring equipment, would obtain the same decimal number and decode the secret message

Assuming that the precision needed for this operation were available, would it work, at least in theory? If not, can you think of a procedure which would make it theoretically viable?

Q07
Tongue-in-Cheek Physics

Air Mail

When posting a parcel, you have to pay more if the parcel is heavier. There is a scale of charges linked to the weight of the parcel. The post office is quite right to charge according to the mass of the parcel. However, I have noticed that the machines used are spring balances, which measure weight rather than mass.

This means that I could employ the following scheme. When I wanted to send an object through the post, I could enclose it in an oversized box, which would leave space for me to put a helium balloon in with it. The parcel has the extra mass of the helium and the fabric of the balloon, but because of the upthrust on the balloon the parcel would weigh less, saving me postage.

Presumably, if the balloon were large enough, the parcel would have negative weight, and the post office would have to pay me to transport it. Would the scheme work?

What machine would correctly measure the mass of the parcel?

Solution on page 218

Q08
Tongue-in-Cheek Physics

The ultimate weapon

D. Stone, in a letter to *Geotimes* (1969), describes a geophysical weapon which, if used by a nation with a large population, could have a devastating effect on an enemy.

Suppose, for instance, that all the residents of China, who now number nearly one billion, were to jump in unison on pogo sticks. Such an action could set up shock waves which would amplify with each jump, perhaps reaching a magnitude of 5 on the Richter scale. This could, for example, destroy parts of the United States. To protect itself, the U.S. might organize jumps timed to cancel the offensive waves. But, considering the difference in populations, the Americans would either need to jump from a greater height or call on its NATO allies for jump assistance.

History would suggest that the Pentagon would probably decide in favour of countering the threat in a manner other than by pogo stick.

However, would this Chinese super-weapon work?

Solution on page 218-19

Q09
Tongue-in-Cheek Physics

Charging for a flight

Michael Faraday (1791-1867) was a British scientist who revolutionised the scientific understanding of electromagnetism, establishing an entirely new field of study. He achieved this despite receiving very little formal education, and consequently having a limited grasp of mathematics.

I know from my school physics classes that whenever a conductor moves through a magnetic field, a voltage is produced across it. This was discovered by Michael Faraday in 1831, and is the basis for all electric generators.

It occurs to me that most aircraft are made of metal, and they move through the Earth's magnetic field. I did a calculation to see how much electricity could be generated. Unfortunately, the Earth's magnetic field is very weak, so even Concorde would generate only about a quarter of one volt between its wingtips. But this might be enough to run some small electronic devices without the need for batteries or generators.

Is this scheme feasible?

Solution on page 219-20

Q10
Tongue-in-Cheek Physics

How do you fly through time?

Most readers will have heard of the International Date Line (IDL), established as an irregular line drawn by convention through the Pacific Ocean, substantially along the 180th meridian. IDL marks the place where the date changes.

Crossing the line from west to east, travellers gain a day and, conversely, lose a day travelling east to west.

The position of the IDL has been arbitrarily designated and is partly curved to accommodate eastern Siberia, then bulges westward again in order to avoid crossing land.

Let us now assume that a successor of Concorde can take its passengers around the Equator, circling the globe in six hours. Flying west to east it will cross the IDL four times in 24 hours and consequently lose four days. Starting, say, on 18 March it will land on 14 March. Continuing the journey the passengers will go back into history, eventually reliving the birth of Christ. Travelling east to west they will travel into the future, giving them a decisive advantage in the lottery on their return.

Is there something wrong with this reasoning?

Solution on page 220

Q11
Tongue-in-Cheek Physics

Feathers and gold 2

Again which weighs more, an ounce of feathers or an ounce of gold?

Q12
Tongue-in-Cheek Physics

How do I make a sound bomb?

Here is a design for a bomb that requires no explosives. Imagine that the bomb was made of a material that was a perfect reflector of sound. Inside it has a spherical cavity. At the centre of this cavity sits a ticking watch. The sound inside will build up, as it cannot escape. Eventually the sound energy would be so great that it could no longer be contained. At this point the bomb would explode.

Would this design work?

Solutions on page 220-221

Solutions
& Explanations

Sight and Sound

How can I turn red to green?

If you filmed the LED with a high-speed camera to slow down the effect, you would see it flash red and green. But our eyes, by a process called "persistence of vision", work much more slowly and you would be unable to see different-coloured flashes.

Also, our eyes do not have a colour receptor for each colour in the rainbow. There are only three types of colour receptor in our eyes, sensitive to mainly red, mainly green, and mainly blue. The other colours are detected by triggering more than one receptor to a greater or lesser extent. Colour televisions make use of this fact; if you look closely at the screen you will see that the picture comprises only red, green and blue dots. Our flashing LED will stimulate the red and the green sensor simultaneously, giving the sensation of yellow light.

How do I dull my headlights?

It would work. However, it was not introduced because the disadvantages would outweigh the benefits.

1. Not being able to see the headlights of approaching cars would be dangerous, particularly when visibility is reduced during heavy rain or fog.
2. The windshield's polarization would absorb a fair amount of light in the street scene, reducing visibility generally.
3. At any rate, it would be difficult to provide an effective filter for windshields in view of their irregular shapes.
4. Coloured stress patterns would become visible in the wind-shield and this would be distracting for the driver.

Can a mirror turn me upside down?

The answers are:
1. Place the mirror on the floor (or ceiling).
2. Place the mirror at a 45° angle to the floor.
3. I cannot think of a way either.

So near and yet so far?

If a telescope (or a pair of binoculars) is being used in the correct way, it is said to be in "normal adjustment". When used in this way, the eye is perfectly relaxed when using it. When a perfect eye is relaxed, it focuses at infinity. A telescope does not, therefore, give you an image that is closer. It does, however, make the object look bigger.

A magnifying glass increases the focusing power of the eye, allowing the object to be placed closer. There are many circumstances in which it works by not magnifying at all. Try

this simple experiment to see what I mean. Hold a piece of writing so close to the eye that it cannot be seen clearly because it is blurred. Now insert a magnifying glass between the eye and the writing. You should now find that the writing is clear enough to read but is no larger than it was.

Why isn't the whole sky as bright as sunlight?

Olbers himself provided the counter-argument.
1. While light from some of the galaxies we observe left them millions of years ago, the light of many millions of stars has not yet reached us. Our own galaxy is more than one hundred thousand light-years across.

2. The paradox assumes that all or most stars in the observable universe are alight at the same time. This is a misconception, because the life span of a star is limited to about 10^{10} years. Although this is a long time, it is still finite.

3. The universe itself is probably not infinite.

4. Light may also be absorbed by dust, and dark matter (if it exists).

Glaringly obvious?

Strangely enough, the answer to this question is "Yes". If there are just two crossed polarizing filters, no light will pass through the combination. If you insert an extra filter between the other two, at 45 degrees to them, then some light will travel through this triple combination. In fact, the amount of light transmitted through the triple combination is about half of what would pass through a single filter.

This is because some light will always manage to get through a double combination at 45 degrees to each other. So some light will pass through the first and second filter. Some light will also pass through the second and third filter. Therefore, some light will pass through the complete triple combination.

Transparent object

Wet ground glass. The commonest example in a kitchen might be a Pyrex baking dish that has become worn or scratched on the bottom so as to be merely translucent, not transparent. When wet, it is clear.

Can you travel faster than light?

My friend correctly identified three situations in which something travels faster than light. However, none of them contradicts Einstein, whose work could be more correctly summarized as: "The maximum speed at which information can travel is represented by the speed of light in a vacuum."

In the first two examples, nothing physical moves faster than light. The barber's pole stripes only appear to move along the pole. The spot of light at one end of the sweep is different from the one at the other end as it is made up of different photons. Neither system could be used to carry information.

The third example has been observed experimentally. Such particles break the "light barrier" and produce minute photonic flashes equivalent to sonic booms.

The spectrum

Pigments in common use are impure. Because of these impurities, yellow paint presents the eye with a mixture of red, yellow, and green, and blue paint offers a mixture of blue and green. In each case, however, the basic colour dominates so that the eye perceives only yellow and only blue respectively. However, when the two paints are blended, yellow paint absorbs blue, and the blue paint absorbs red and yellow light, leaving green as the only remaining colour common to both paints.

Why can't I tune my radio?

There would be even worse problems if two or more transmitters attempted to broadcast the same programme on the same frequency.

As radio is a type of wave, at some points the two waves from the two transmitters could just happen to be synchronized so the radio would pick up a very strong signal. At other places the waves could be completely out of synch and would cancel each other out. Thus, as you moved around, the radio would continually go from extremely loud to no signal at all within the distance of a few metres.

Can you see 3D?

The phenomenon of 3D vision rests on the fact that the two eyes perceive slightly different versions of the same object, as shown in the illustration.

The human brain is genetically programmed to integrate the two views of the same objects into three-dimensional vision. If you close one eye, the three-dimensional world you see around you becomes two-dimensional. You might not see much difference, but that is because your visual memory knows that what you are looking at is, in fact, three-dimensional and persists in "seeing it" that way. A one-eyed person, however, is unable to see in three dimensions so cannot estimate distances, an ability that relies on the processing of different images received by the two eyes.

Any attempt to make a two-dimensional image appear three-dimensional relies on deceiving the brain into reacting as if each eye were receiving a different image of the same object.

Years ago, the stereoscope, an optical instrument with a binocular eyepiece, presented two slightly different pictures or photographs for viewing, one with each eye. The brain was thus deceived and offered the illusion of a three-dimensional picture. The short-lived 3D movie presented overlapping images in red and green that were filtered out by the red- and green-lensed spectacles, again fooling the brain. These 3D illusions, and some others, are based on the fact that human eyes usually focus on whatever object they are looking at. However, the eyes can defocus, or diverge, and this is the principle behind stereograms.

Black to the future

White is, strictly speaking, not a colour but the presence of all colours. A white bulb gives out all visible colours. Black, also, is not a colour but the complete absence of all light. By definition, therefore, there can be no such thing as a source of "black light".

Can you stop an echo?

No recognizable echo, which is a reflection of sound, but the decibels (intensity) would be doubled, much as is candlelight when placed next to a mirror.

Rays of the Sun

The Sun's rays are parallel, of course. The "fanning out" is an optical illusion, the same illusion that makes railroad tracks appear to meet at a point on the horizon. These rays are a more powerful illusion because of the absence of reference points. It's hard to believe, even when you know it for a fact, that two rays high overhead and "so far apart" are just as far apart in the distance where they first "emerge" from behind the cloud.

The well-trained sunglasses

Early photochromic sunglasses contained tiny crystals of silver chemicals known as halides in the glass. These are transparent, but when hit by ultraviolet light, the silver ions combine with electrons to form silver atons and halogen gas. The gas cannot escape the glass, and the silver atoms clump together into specks of silver, darkening the glasses. Away from the light, the silver atoms can no longer hold the extra electrons, and recombine with the gas, returning to the transparent silver halide form. Modern glasses contain carbon-based molecules which get bigger and darker when warmed up by ultraviolet, and smaller and lighter when they cool down.

How do I blow up my stereo?

Even with correctly connected speakers you find that in most parts of the room there will be several frequencies of sound that combine destructively while others combine constructively. This means that some frequencies will be quieter than they should be, while others are louder; but this is not normally noticeable among the thousands of other frequencies present. At any point equidistant from each speaker, however, you can guarantee that all the frequencies will combine constructively, giving the best listening position.

But if you wire one of the speakers with the reverse polarity, at the central position all frequencies combine destructively, giving the worst possible listening situation.

What's in the middle of a rainbow?

At the circle's centre is the shadow of your head. The droplets in a rainbow are on the surface of an imaginary cone that points straight to the Sun, behind you, and has its vertex at your eyes. (In the case of nearby rainbows, as in a lawn sprinkler, you may be able to see twin, overlapping bows, one from each eye.) A rainbow's arch is always 42 degrees away from the line of sight between you and your shadow. The full circle fills a visual angle of 84 degrees, nearly a right angle.

How did we avoid rockets?

The V-2 rockets exceeded the speed of sound and therefore you could not hear their approach. You could hear only the detonation after they hit the target, too late to take evasive measures.

Light fantastic

In one of the rooms, the light is provided by sodium lights, which provide a very pure yellow light. In the other room the light is provided by a mixture of red and green lights. Red and green lights, when mixed, appear yellow. In the first room, both red and green paint appear black under the pure sodium light, while in the second room the red and green paints are able to reflect their own colours.

Can you fly faster than sound?

1. There is a common misconception that a sonic boom occurs only at the moment when a plane exceeds the speed of sound. In fact, any aircraft travelling faster than sound creates a large pressure disturbance that travels along with the plane. If this pressure disturbance passes a person on the ground it is detected as an explosive sound. That is why the Concorde is not allowed to fly supersonically over land. As this pressure wave does not pass the passengers inside the plane, they are unaware of the sonic boom.

2. No. Breaking the sound barrier is a common expression meaning travelling faster than sound. Sound does not always travel at the same speed; it is affected by the air temperature. As air is colder at high altitudes, the speed of sound is correspondingly lower by about half a metre per degree of Celsius. At 0°C, the speed of sound is about 328 metres per second, and at 16°C about 338 metres per second.

What makes a mirror reflect?

Left and right are directional concepts while top and bottom, or up and down, are positional concepts. The same incidentally applies to east-west and north-south. Walk northward along the Greenwich Meridian and Berlin will be to the east and on your right (ignoring the fact that you could also travel to Berlin the long way round) until you reach the North Pole. Crossing it, you turn around to still look north and Berlin will be to the

west and on your left. Yet north and south will remain in the direction of the poles. Equally, up and down assume the centre of the Earth as reference point. An ordinary mirror will reverse direction, but not position.

Why do you have two eyes?

With two ears you can identify the direction from which a sound is coming. The same principle is used in direction-finding equipment to locate the source of illegal radio transmissions. With only one eye you would be unable to see objects in three dimensions, nor would you be able to estimate distances accurately.

What makes a sunset's colours?

A number of books have been written on the subject. In simple terms, the vivid colouring of sunset is due to the scattering of sunlight by atmospheric dust, whereby the shorter blue light of the spectrum is scattered more than the longer wavelengths of the red end. Dust from many sources, for instance volcanoes, contributes to the variety of colour display.

For once, I would suggest that you forget physics and continue to enjoy the unique experience of a glorious sunset.

Why doesn't my kettle make a noise?

Where the water is being heated it is locally hotter than elsewhere. Small bubbles of water vapour will form, which collapse as they move to cooler parts. The "singing" sound is caused by these thousands of tiny collapsing bubbles. When the water is close to boiling, the bubbles survive, causing the kettle to go quiet.

How do you magnify glass?

Shortsightedness (myopia) is caused by defective refraction in the eye's lens. Near objects can be seen clearly while objects in the distance appear blurred because they are focused in front of the retina. Concave lenses change refraction and correct the defect.

Strangely enough, pinholes have a similar effect. Therefore, all you have to do is find a piece of cardboard and make a pinhole. Looking through it, you will be able to read the number plate.

Why does the ringing change?

There was a strip of railing nearby. Sound reflected back from each individual support rail arrived as a series of small echoes separated by a small interval of time. This caused a distinct ringing tone, the frequency of which was determined by the distance between the rails.

Death fall

The sound effect of someone falling ordinarily gets fainter in films, but it does not lower in pitch as it should according to the Doppler effect – as the faller's speed increases (acceleration), the sound frequency lessens.

Sound reasoning

Although light and sound are both wave forms, there are some significant differences, the major one being wavelengths. Sound waves are typically one metre in length, whereas the wavelength of light is about one ten-thousandth of a millimetre. Also, sound is 'coherent' (continuous) whereas all light sources (except lasers) produce light in the form of millions of small packets which are all jumbled up.

1. Waves will curl around obstacles (diffract) when their size is of the same order as the wavelength. Thus sound will easily diffract around buildings, trees, etc. This can also be seen for light, but the obstacles have to be much smaller. For instance, a diffraction pattern can often be seen while looking at a bright point of light through a closely woven material.
2. Light can also be made to combine destructively, just like sound, but because of the small wavelength and lack of coherence this can be difficult to observe. This is what causes the characteristic "speckled" appearance of a spot of laser light and the colours seen in soap bubbles and films of oil.

Why are clear things coloured?

There is no substance that is perfectly transparent. Light is gradually absorbed as it passes through an increasing thickness of any substance. That is why it is completely dark at the bottom of the ocean.

If all colours are absorbed equally, there is no resultant colour. Ordinary glass has a slight tendency to absorb magenta, leaving the green hue seen through thick glass.

The sky appears blue because that colour has the shortest wave length and it is scattered by small dust particles in the air.

There can be two reasons why mountain lakes look blue. First, the water is reflecting light from the sky. Second, water from melted glaciers is often coloured because of the high mineral content.

Space and Celestial Bodies

Can you square a circle?

The apparent gravity increases the farther one moves from the point of rotation. The gravitational effect would therefore be greatest at the edges of the room. For example, a marble placed in the centre of the floor would roll towards one of the two edges, which are parallel to the axis of rotation.

Even though the floor is perfectly flat, it would feel as though there were a hump in the room, that one would be walking "uphill" towards its centre.

Why aren't I floating?

There is no way to distinguish between 1) gravity and 2) acceleration. However, in a rotating spacecraft "gravity" will appear greater farther away from the centre of rotation, so there are several things that he could do:

1. Weighing an object with a spring balance will give a smaller reading near the ceiling.
2. A marble placed in the centre of the floor will roll towards the edge.
3. Two plumb lines will diverge rather than hang parallel.
4. He would not be able to spin a coin.

Why do I fly back to Earth?

At first the atmospheric drag is very small and has the effect of making the satellite fall into a lower orbit. The speed in orbit is determined by height: the lower the orbit the faster the speed. The satellite therefore travels faster.

This may seem like a puzzling contradiction to the law of conservation of energy. But the total energy of the satellite, potential and kinetic, has in fact decreased by the amount of heat produced by friction.

Eventually, as the orbit spirals downwards, the frictional drag will be great enough to slow down the satellite as it falls into the main body of the atmosphere. The increase in orbital speed is similar to the effect on a skater who pulls his outstretched arms nearer to his body, thereby increasing his speed of rotation.

Can you propel a boat with a rope?

By jerking on a rope attached to its stern, a small boat can indeed be moved forward in still water, and speeds of several miles per hour can be achieved. As the person's body moves toward the bow, the friction between the boat and the water prevents any significant backwards movement of the boat, but the

inertial force of the jerk is strong enough to overcome the resistance of the water and transmit a forward impulse to the boat.

The space capsule, in the absence of friction, cannot be propelled in the same way.

How long to travel through space?

It would seem that we can get there in a minimum of 4.3 years, but we must instead determine the maximum acceleration and deceleration a human body can stand to reach and return from a speed which has been approximated at 186,000 miles per second.

Are there rainbows on the Moon?

Part of the mysterious beauty of the rainbow:

1. You can see the spectrum only if the angle of refraction between the Sun, the drop of water, and your line of vision is between 40° and 42°.
2. Lunar rainbows are also possible but rare, because moonlight is not as strong as sunlight and its intensity varies with the phases of the Moon.
3. This phenomenon is called dew-bow and is caused by the water drops on the grass.

Why does the Sun go out?

There are two reasons why solar eclipses are so rare:

1. The Moon's orbit is not in the same plane as the Earth's orbit around the Sun. This means that every month when the Moon is on the "right side" it is usually too far to the north or to the south to block the Sun's light.
2. If the two orbits were in the same plane, a total eclipse would occur every month, but it would still be seen from very few places on Earth. This is because the Moon's shadow, when cast on the Earth, is only a few miles across. Only if observers happen to be in this shadow area will they experience a total eclipse.

The space station

1. If the astronaut walked round the station in the direction of its rotation his weight would increase. If he turned round and walked in the opposite direction his weight would decrease. Why should this be? We know that the centripetal force increases proportionally with the speed of rotation. By walking in that direction the astronaut adds his speed to that of the rotation.
2. Try to spin one of your coins on the floor of your room: the coin will refuse to spin. By conservation of angular momentum, a

spinning object tries to maintain its position in space. Since the spinning station is continually changing your position in space, a coin that is spun will keep changing its orientation, to correct its angular momentum, and will topple and fall.

Pull of the Sun

Let us consider a simpler system first. Suppose there were only the Earth and the Sun. If the Earth were stationary in space, the two would be drawn together by their mutual gravitational attraction. However, there is a more stable possibility: the Sun's gravitational force is used to maintain the Earth's orbit rather than drawing it closer.

The same process explains why the Moon orbits the Earth. The Sun's gravity is used to maintain the Earth/Moon system in orbit. If all orbiting suddenly ceased, all the planets and their moons would instead fall towards the Sun.

What colour Is the Sun?

I can think of five answers:
1. white – almost by definition, the colour of sunlight is white.
2. all colours – one could argue that there is no such "colour" as white. Isaac Newton discovered that white light is really a mixture of all visible colours.
3. yellow – the most abundant colour in sunlight.

4. black – the Sun appears white because it is emitting light, rather in the way a light bulb looks white while it is on; the bulb element is grey when the bulb is off. But what colour would the Sun be if it cooled down? Scientists know that the Sun is close to being what is called a Black Body Radiator. So if it cooled down it would be black.
5. no colour at all – black is not really a colour but is the absence of all colours.

Does the Earth move?

There are two ways in which earthquake waves can travel around the world. First, they can travel across the Earth's surface, rather like waves travel across the sea. Second, they can travel directly through the centre of the Earth. These waves travel at different speeds and have different distances to travel, so they arrive at the seismograph at different times.

The Earth does not have a uniform density. The denser core acts like a lens, focusing the vibrations at some points while leaving other places in a "shadow".

Drops and bubbles

The bubbles would attract each other. If water is removed from one spot in all space (bubble A), the gravitational balance surrounding it is upset, and the net effect on a nearby molecule

of water is that it is drawn toward greater mass; that is outwards, away from the bubble. If there are two bubbles, the water between them acts as if it is repelled from both, and the bubbles would move towards each other.

How does the Moon cause tides?

It is convenient to speak of the Moon going around the Earth, but it is more accurate to say that they form a two-body system that revolves around its centre. This common centre of gravity is actually inside the Earth because of the Earth's greater mass. As the Earth swings around this centre, centrifugal force causes the ocean waters to flow away from the centre, producing two outward bulges on opposite sides of the globe.

How do you photograph the Moon?

Photographs show that the Moon is the same size in both positions. The illusion is universal – it is apparent even in a planetarium, but there is still no fully accepted explanation for it. Ptolemy argued that the horizon Moon appears larger because we can compare it with distant trees and buildings. This theory is still the most widely accepted one, but it doesn't explain why sailors see the Moon illusion just as vividly at sea.

Can you see all of the Moon?

No, it is not a coincidence. It is well known that the Moon exerts tidal forces on the Earth's oceans. What is less well known is that there is also a tidal effect on the Earth's solid surface. The total result of this is to slow the rotation of the Earth, each day being slightly longer than the preceding one. This is why occasionally a "leap second" has to be added to the length of a day to keep clocks accurate.

Many scientists believe that at some time in the past the Moon was rotating faster, and that tidal forces in its surface slowed its rotation until it became synchronous. This phenomenon is quite common in the solar system; for instance, Titan takes 15 days and 23 hours to orbit Saturn and to rotate once about its own axis.

Other cosmologists take a different view. Peter Goldreich, in an article in *Scientific American*, believes that the synchronous rotation of the Moon is caused by the fact that the Moon's mass is not uniformly distributed and that the Earth's gravitational field acts upon the heavier half of the Moon.

Ring around

If a relatively weak material were used, for example string, the ring would collapse to the ground. If the material were very rigid, it would be in unstable equilibrium, so would touch the ground at one point.

The ring could be made stable if it were rotating, because of the even outward force generated, but air friction would soon bring it to rest.

How fast can a space rocket travel?

The car stops accelerating when the thrust of its engine equals the drag of resistance or friction. A rocket in outer space experiences no friction, so there is no limit to its acceleration. However, as things get faster their mass increases, making it more difficult to accellerate them; so the rocket will never exceed the speed of light.

The spy satellite

The orbit of a satellite must be around the centre of the Earth, which is the centre of its gravitational field. There is no way in which a satellite can be synchronous with Baghdad's latitude. It follows, therefore, that a satellite can hover only over a point on the Equator.

How do you relay information in space?

The time it takes for an object to orbit the Sun is determined only by its distance from it. The Earth takes 365.3 days to orbit the Sun because it is 93 million miles from it. At a position halfway to the Sun, the period of rotation can be calculated as 130 days. The satellite could not orbit the Sun and maintain a movement in conjunction with the period of orbit of the Earth.

Shape of the flame

The characteristic shape of a candle flame is caused by convection as the hot exhaust gases are pushed up by the colder, fresher air.

In weightless conditions there would be no convection as there is no up and down. If there were no drafts the flame would be spherical, but would not last very long because local oxygen would soon be used up, with no air currents to replenish it.

Moonshine

A dark part of the Moon, which is not directly illuminated by the Sun, nevertheless receives some sunlight indirectly, which is reflected from the Earth's surface.

Can you steer in space with a shotgun?

Firing the bullets will propel you by recoil towards the spaceship, if fired in the opposite direction. With each bullet your speed will accelerate.

Suppose the bullets leave the barrel at 1,000 metres per second. Then, depending on the ratio of the mass between the bullet and you, the acceleration could be as much or more than free fall in the Earth's gravitational field (about 10 metres per second2) and you would not survive the impact with the craft. You would have to decelerate by firing the last bullets in the direction of the spaceship, in order to reduce the velocity with which you hit the spacecraft. You might be better not actually shooting the hull, however.

The asteroid

Zero gravity would prevail at all points inside the hollowed-out asteroid, so a floating object would maintain its position.

How high can I jump on the Moon?

About 5.6 metres (18.5) feet. It isn't the distance from the ground to the bar that is important, but the distance from the jumper's centre of gravity (CoG) to the bar. A 180-cm tall (6-ft) male's CoG is about 1m (3.5 ft) above the ground; so to jump his height he must raise his CoG 75cm (2.5 ft). On the Moon he would be able to raise his CoG six times as high, or 4.5 m (15 ft). By raising his legs up to clear the bar the same way he does on Earth, he could add that metre back to the jump (just as on Earth), bringing him to the 5.6 m mark (18 ft 5 ins) mark, which is just over half the height that is commonly quoted.

Does the Earth revolve around the Sun?

Experiment 1: Weigh yourself on the Equator and then on one of the poles. You will weigh slightly less on the Equator because the centrifugal force acts against gravity. This is not the case at the poles.

Experiment 2: Use a train travelling in the northern hemisphere towards the North Pole, starting from the Equator. If, as we want to prove, the Earth rotates, then the easterly velocity will be maximum at the Equator, decreasing as we travel north, becoming zero at the North Pole. But the train itself would try to retain the maximum easterly velocity by virtue of the phenomenon of inertia. The rails prevent the easterly motion, and because of this the right-hand rail is subject to increased pressure, which could be measured.

Experiment 3: Drop an object into a deep mine shaft at the North Pole and Equator. Like the train, it will have easterly deviation at the Equator. With a depth of 100 metres, the easterly deviation is 2.2 cm at the Equator.

I hope you are now satisfied. The effect of the Earth's rotation is also evident in the phenomena of trade winds and ocean currents.

Why do you get more meteors in the morning?

As the Earth passes through a swarm of meteors its "face" – that is, in the direction the Earth is travelling around the Sun – encounters more meteors than its "back" does. For a similar reason, when you run through the rain more water falls on your face than on the back of your head. In the early evening, you are on the trailing side of the Earth and the only meteors you will see are those that "overtake" the Earth from behind. Between midnight and sunrise, you are on the Earth's face and are looking in the direction from which most collisions come.

The phenomenon is the same in Australia and everywhere else on the planet.

Why are there two tides per day?

It is convenient to speak of the Moon going around the Earth, but it is more accurate to say that they form a two-body system that revolves around its centre. This common centre of gravity is actually inside the Earth because of the Earth's greater mass. As the Earth swings around this centre, centrifugal force causes the ocean waters to flow away from the centre, producing two outward bulges on opposite sides of the globe.

Hot and Cold

How do I make a toy boat?

As the steam enters the cooler tubes it condenses, creating a vacuum that will draw in water. However, such water as is sucked in will not do so in the form of a jet, but will come from all directions and will not therefore have the same propulsive force.

Why is the safety lamp safe?

The explosive gases would still infiltrate through the mesh and ignite inside the lamp. The screen would quickly conduct away the heat, thereby preventing the flame from escaping to cause an explosion. The change in the flame also gives a warning of danger to the miner.

Why don't I burn my head?

I had dried my ears with a towel and therefore they felt the real temperature of the air. My hair was still wet, so the fast-moving air was rapidly evaporating the water. Evaporation produces cooling (this is the principle used in the domestic refrigerator), and therefore the air felt cool on the wet hair.

How do you drink on ice?

Polar ice does indeed contain salt. Such ice, melted down, is as undrinkable as seawater. However, over time, the brine in the ice blocks will migrate downwards, because of gravity. This draining effect will make the melted ice drinkable after about a year, and it will be almost completely free of salt after several years.

The problem does not affect all polar regions, as some ice is formed by precipitation. However, many areas are arid, with insignificant snowfall, and such little as there is will be blown away by winds of up to 100 mph.

Climbing the mountain

Bob was most correct when he stated that the cold was due to the rarefied atmosphere. While it is true that the sun's rays are stronger at greater heights and that winds have a cooling effect, and the greater the height the stronger the winds, the main reason for the lower temperature at greater heights is the rarefied atmosphere. With fewer molecules per cubic inch, fewer molecules bombard us each second and less heat is transmitted to us by the surrounding body of air.

How do you drive on ice?

Good drivers know:

1. To increase friction where it matters you will use the blanket in front for a front-wheel-drive vehicle, otherwise in the rear.
2. The lower the torque the better; therefore, you should start in second gear at slow speed.
3. To stop skidding you should turn your front wheels into the skid.

Why doesn't the snow fall evenly?

The wind driving the snow diverges many metres in front of a large building, thus dispersing the snow before it hits the wind-side face. A smaller object does not divert the wind, permitting the snow to build up.

Does wind affect my mind?

There are three questions to be answered:
1. Why is the wind warm? The meteorological conditions have to be right for the cold mountain air to sink in the form of a fast-moving wind. As the air sinks it encounters the higher pressures found at lower altitudes and is compressed. This compression causes warming; the result of a phenomenon known as adiabatic process (see *Glossary* on page 220).
2. Why is the wind dry? The air was originally cold and cold air can hold less moisture than hot air.
3. Why does it have psychological effects? Nobody is sure that it does, or if so what the reason might be.

Why is Death Valley so hot?

A number of factors contribute to these extreme conditions. The mountain range on the west side rising to more than 3,300 metres absorbs the moisture of the west winds, which, on descending east of the Rockies, are adiabatically heated and dried, turning the valley into a hot, arid desert.

How do you cool off?

You might cool off temporarily, but the net effect of leaving the refrigerator door open will be to make the room hotter. The refrigerator's cooling system will activate in an attempt to cool the refrigerator air again, and more heat will be released by the motor than can be absorbed by the released cool air, so the apartment will become even hotter.

Today's New Yorkers are only too aware of this effect, complaining that, in summer, subway platforms seem to be even hotter than they used to be. Since air-conditioned subway cars were introduced, they are! The air-conditioning systems' motors give off more heat when a train is in a station than the cooling effect of the cold air escaping through open doors.

Why does my tap steam?

The answer is so simple that I should have known the first time I noticed. The hot water heats the tap valve first, which expands, reducing the flow. After the whole tap has warmed up, the effect disappears.

Thermal expansion

The initial fall to level B-B is due to the expansion of the flask, which is heated and expands before the heat has had time to act upon the water. Once the glass of the flask has

reached the limit of its expansion, the water in the tube begins to rise.

Why is the floor cold?

The two floors were probably at exactly the same temperature. However, carpet is a much poorer conductor of heat than are the tiles. As the tiles were taking heat from my feet faster, the tiling felt colder than the carpet.

Breath on your hand

There are two possible explanations involving: a) evaporation and b) expansion.

The blown air moves more rapidly, causing moisture in the skin to evaporate quickly, which produces a cooling effect. Also, as the air passes through pursed lips, it expands and therefore cools. This is the opposite effect to air being compressed in a bicycle pump becoming warm.

You can prove that both effects are present with a piece of clear food-wrap. If you place this over your hand, preventing evaporation, the blown air still feels slightly cold. Wetting the outside of the food wrap allows evaporation to occur, restoring the full cooling effect.

Blowing out the candle

The trick is to lower or raise the funnel until the flame of the candle is in line with the upper or lower dotted line. Any attempt to blow out the candle with the flame opposite the centre

of the opening of the funnel is hopeless, because the movement of air is in the boundry layer.

How do I boil water in my hand?

The experiment is not a sleight of hand. It works! Keeping the lid on the pan slightly increases the pressure, causing the water to boil at a higher temperature (salting the water also increases the boiling point of the water). Because the temperature is higher, the food cooks more quickly.

Insulating metal pipes

The pipe will lose heat by three processes: conduction, convection, and radiation. The amount of heat lost through radiation depends on the excess temperature of the surface and the total surface area. The insulation will reduce the surface temperature, although it also increases the surface area. As we have seen, the insulator may well increase the heat lost through conduction.

However, most of the heat is probably lost through convection. The amount of this will be determined by the surface temperature, which is greatly reduced by the insulating material. If there were no insulation, and we just relied on using a thicker pipe, this would have very little

effect on the surface temperature but would increase the surface area. This would increase the heat lost through convection and radiation.

Your biggest fan

It circulates the hot air that stoves give off. Fireplaces (non-smoking ones) only give off radiant heat, which is unblowable.

How do you walk on fire?

Indeed physics can explain these phenomena. The tongue is obviously wet and the feet, even if not specially prepared, are moist. The moisture in contact with the red-hot substance is immediately vaporized, forming a protective cushion of gas around the feet or tongue.

Some people wet their finger and then dab it on a hot iron to check on its temperature. If done quickly, this does not hurt, and they can tell from the sound how hot it is. Although this explanation solves the mystery, readers are strongly advised not to try – it's very easy to get it wrong and injure yourself if you don't know exactly what you're doing.

Iron doughnut

As the doughnut expands, it keeps its same proportions; so the hole also gets bigger. This same principle is at work when an optician removes a lens from a pair of glasses by heating the frame. The next time you can't open the metal lid on a stubborn jar, heat the lid under hot water. The lid, inner circumference and all, will expand, making it easier to loosen.

How do you open a decanter?

You will likely need someone's help with this: another pair of hands to hold the bottle firmly while you wrap a narrow strip of cloth around the bottle's neck. Then, by pulling the cloth rapidly to and fro, you can cause friction between the strip and the glass that generates sufficient heat to expand only the bottle's neck so that the stopper can be withdrawn easily.

Heating otherwise does not do the trick, as bottle and stopper would expand uniformly.

Can you start a fire with a mirror?

Archimedes' feat is entirely practical. It was reconstructed in 1973 by a Greek engineer who had 70 mirrors (each about 5 feet x 3 feet) held by soldiers, who focused the Sun's rays on a rowboat that was anchored about 50 metres (160ft) offshore. A few seconds after the mirrors had been properly aimed, the boat started burning, and was eventually engulfed in flames. In order to work, the mirrors would have to be slightly concave, with the focus on the rowboat.

Weighty

How can I balance my scales?

You might think that the correct result is the average, 122 grams, but this would be true only for a balance with equal arms. This balance must have unequal arms, and the formula for this is the square root of the product: 100 x 144 = 120 grams

Proof: If the scale's balance when empty is inaccurate, the two arms must be of different lengths (a, b).

Let the corresponding readings for the object X be W1 and W2.

At balance:
aX = bW 1 -> a/b = W 1/X and aW2 = bX -> a/b = X/W2
Therefore:
W1/X = X/W2 or
X = SqRt (W1 x W2)
Substituting:
X = SqRt (144 x 100) = 120 grams.

Quito

He would weigh zero. His centre of gravity would be in orbit, considering the Earth's speed of rotation and the fact that Quito is on the equator.

How can I make a spray gun?

The air travelling through the blow tube at considerable speed reduces the air pressure in the conduit tube. The liquid in the container is subject to atmospheric pressure and is therefore forced up, and out of, the conduit tube in the form of a fine spray.

Why don't submarines rest?

Water pressure pushes perpendicular to a submarine's hull at every point, at the bottom as well as at the top and sides. When a sub settles on a clay or sandy bottom, the water layer may be squeezed out from beneath the hull, robbing the sub of much of its upward buoyant force. In effect, the downward forces can then "glue" the sub to the bottom.

How do you steer a boat?

The rudder will have an effect only if there is relative motion between the boat and the water. This answer begs the question 'Is there such relative motion?' Most likely yes. However, there are so many forces acting upon the boat that this question cannot be answered with absolute certainty.

First, there is gravity. Imagine the river to be frozen. Ignoring friction, the boat would

slide down the slope. Then there is buoyancy, resulting in a component force downriver, partly counteracted by drag. Air resistance adds to the drag, while a wind upstream or downstream has an effect. Furthermore, the river flows at different rates in the middle and near the banks.

It is theoretically possible that all these forces combine to synchronize the motions of the boat and river for a limited period, which would make steering impossible.

Balloon in a bath

It stays where you move it to –
go on, give it a try!

How do you walk on a tightrope?

Walking a tightrope is possible only if you keep your centre of gravity precisely above the rope. This is a bit like keeping upright on a stationary bicycle.

The balancing pole is deliberately heavy, with weights placed in the tips. It makes things easier in two different ways. The bar has a large inertia, and by moving it sideways the walker can adjust his position. Also because the bar is heavy at the ends and bends downwards, this has the effect of lowering the centre of gravity of the walker. If this could be lowered to below the rope, then the walker, although still looking very precarious, would in fact be very stable.

Does a bootstrap elevator work?

Although it looks as if the man were trying to lift himself up by his own bootstraps, he really isn't. True, for every pound of force that he pulls up on the rope, he also pushes down an equal force on the block, but if he is strong enough to lift his own weight plus the weight of the block, he will rise from the ground. (Tests have shown that a 190-pound man can lift both himself and a 110-pound block in this way.)

Gravity train

The gravity train would indeed work. Strangely enough, such a train would make all of its trips in about 42 minutes, the same time it would take an object to fall through the centre of the Earth. The time is constant regardless of the tunnel's length.

Hourglass puzzle

When the sand is in the top compartment of the hourglass, the high centre of gravity tips the hourglass to one side. It is kept at the bottom of the cylinder by the resulting friction against the side of the cylinder. After sufficient sand has flowed into the lower compartment to make the hourglass float upright, the loss of friction enables it to rise to the top of the cylinder.

It is interesting to note that, if the hourglass is a little heavier than the water it displaces,

the apparatus works in reverse. That is, the hourglass normally rests at the bottom of the cylinder, and when the cylinder is turned upside down, it stays at the top, sinking only after sufficient sand has flowed to eliminate the friction.

The toy is said to have been invented by a Czechoslovakian glassblower, who made them in a shop just outside Paris. Strangely enough, physicists seem to find it more puzzling than other people. They often advance explanations which involve the force of the falling sand keeping the hourglass at the bottom. However, it is easy to show that the net weight of the hourglass remains the same whether or not the sand is pouring.

The half-hidden balance 1

From one pound to infinity. It is 1 pound, not 2, because when the length approaches infinity the second weight converges to zero. When the hidden rod's length is zero the weight must be infinite to balance the 1 foot–pound force on the left.

Hole through the Earth

1. The ball's velocity would steadily increase from zero at point A to a maximum at the centre of the Earth. It would steadily decrease thereafter to zero at point B, taking 42 minutes for the complete trip. This fascinating speculation goes back to Plutarch. Even Francis Bacon and Voltaire argued about it. Galileo gave the authoritative answer which is generally accepted: The ball would fall faster and faster, though with decreasing rate of acceleration, until it reached maximum velocity, about 5 miles per second, at the Earth's centre. It would then decelerate until its speed reached zero at the far end of the hole. If air resistance is ignored, it would oscillate back and forth, like a pendulum, *ad infinitum.*

2. At the centre of the Earth, there is no overall gravitational force on the ball. This is because the Earth's pull on the ball will he equal in all directions. Therefore, the weight will be zero.
3. The weight will change but the mass will not.
4. The ball would be in free-fall throughout the entire trip, so it would always be in a state of zero gravity.
5. More time. The trip would take about 53 minutes. Although the distance is much shorter than on Earth, the moon's gravity is only about one-sixth of that of the Earth.

Rope and pulley

The answer is that, regardless of how the man climbs – fast, slow or by jumps – the man and weight always remain opposite. The man cannot get above or below the weight, even by letting go of the rope, dropping, and grabbing the rope again.

The tank

If the tank is suddenly moved to the right, inertia will cause the steel sphere to move to the left relative to the tank. The steel sphere tends to persist in its state of rest. The water tends to do the same, but the sphere, being heavier than water, dominates. Because the cork sphere is lighter than water, it moves to the right relative to the tank. If the tank were moved back and forth, the two spheres would also move back and forth: the steel sphere in the opposite direction, the cork sphere in the same direction.

The bottle and the coin

Isaac Newton's first law of motion explains this experiment:

Every body continues in its state of rest or of uniform motion in a straight line, unless compelled by some external force to act otherwise.

This tendency to continue in its state of rest or uniform motion is called 'inertia.'

Several other tricks are based on the same principle. For example, if a pile of coins is placed on a table, the bottom one can be removed without disturbing the remainder simply by flicking it sharply with a piece of thin wood or metal. In both cases, the incidence of friction is insufficient to overcome inertia.

Fishy business

Let us consider the syringe first. Drawing some air into the syringe will increase the weight of the system because of the extra weight of the air. But at the same time the piston in the syringe moves back. The water level rises slightly, increasing the volume of the system, thus increasing the upthrust exerted on it by the atmosphere. By the Archimedes Principle, the increase in upthrust will equal the weight of the extra air. So there will be no overall change in weight.

The same would be true for the fish, so long as the drawn air was not compressed or absorbed. As both those things are likely to happen in the fish, the change in upthrust will be less than the weight increase. Therefore, the overall weight of the system will increase.

What does an hourglass weigh?

The weight of the hourglass will not change, although some of the grains of sand will be weightless, being in free fall. This will be balanced by the impact force when the grains hit the bottom. There may be some slight fluctuation at the beginning and end, however.

The half-hidden balance 2

The answer is that 2 cubic centimetres of water must be added to the left-hand beaker: one to balance the cubic centimetre of water added to the right-hand beaker and one to balance the buoyant force exerted on the iron bar due to its displacing one cubic centimetre of water when the water level has been raised one centimetre.

How do you skim stones?

The first skip is short and to the right. When the stone's trailing edge hits, it pushes sand to the left; the stone tilts forward and hops

to the right. Then the leading edge strikes and pushes sand to the right; the stone tilts back and takes a long hop to the left, and the cycle repeats. The short hops appear to be missing when stones skip over water. After the trailing edge strikes, the stone planes along, building a crest of water in front of it, then lifts out and makes a long hop. It strikes with its trailing edge again and repeats the process.

Pendulum clocks

No change. Everything falls at the same speed, and gravity is the sole cause of the pendulum's downward motion. This is an important phenomenon in physics.

Galileo Galilei was the first scientist to recognize this principle. He is particularly remembered for the work he did on the acceleration of falling bodies. The Greek philosopher Aristotle had taught that the speed with which a body fell to the ground depended on its mass. This incorrect assertion had been accepted for centuries, but legend has it that Galileo tested the matter by a very simple experiment. He ascended the Leaning Tower of Pisa and, from the top, simultaneously released three iron balls of different masses. They all reached the ground at the same time.

It is very easy to draw a wrong conclusion from casual observation. A feather, for example, falls much more slowly than a stone. The fact is that air resistance as well as force of gravity govern the rate at which a body falls towards the Earth. In the case of a light feather of large surface area, the air resistance is very great compared to the force of gravity on it. If

air resistance is eliminated the feather falls with the same acceleration as the stone. This was first demonstrated, shortly after Galileo's death, by Robert Boyle. Using his newly invented air pump, Boyle removed the air from a glass jar containing a lead bullet and a feather. When the jar was inverted both bullet and feather reached the bottom simultaneously.

How do you burst a barrel?

The trick works because the water pressure at one point depends on depth, not the weight of water above. If the tube connected to the top of the barrel is narrow, a small amount of water can cause a large increase of pressure in the barrel by increasing its depth below the surface.

If a narrower tube were used, then even less water would be needed to increase the depth. If the tube were half the area of cross section, then half the amount of water would be needed to attain the same height in the tube. So half the jug would be sufficient.

The wheelbarrow

Pulling a wheelbarrow is easier. Pushing presses the wheel harder against the ground, adding to the workload.

A question of stability

Item a is a cone, b is a sphere of material heavier than air, c is a dumbbell, and d is a sphere with an off-centre centre of gravity. If the only equilibria an object has are unstable, with no stable or neutral position, a small perturbation would have a most dramatic

effect. Inertia would ensure it would be unable to reach another equilibrium position naturally, and would attain perpetual motion.

What's in the box?

The first box contains uncooked rice piled at an angle. The second box contains an iron weight and a pile of ice. It has to be finely balanced, so that the weight of the ice keeps the box on the table. In due course the ice will melt, toppling the box.

How do you balance a ruler?

The finger that slides first is the one with less friction between it and the ruler. The friction depends on the weight of the ruler on the finger.

How can I get off the table?

If I throw an object in a horizontal direction, the common centre of gravity of myself and this object will still remain on the same vertical line. Thus, if I throw the object, for instance my jacket, to the right, my centre of gravity will move to the left, although more slowly than the object, assuming it has smaller mass. Slowly I will slide off the table, even if I only exhale air by blowing. In this case I would effectively act like a rocket.

The perforated water can

The three answers are:
1. The pressure in the can is less than the pressure outside. The inside pressure is caused by a few centimetres of water while the outside pressure is caused by the atmosphere, which is equivalent to about 10 metres of water. Water will not flow from a low- to a high-pressure area, as that would be like flowing uphill.
2. By drilling a hole in the top of the lid, the

atmospheric pressure is applied to the surface of the water; the pressure outside is still atmospheric, but the pressure inside is now atmospheric plus a few centimetres of water. The water therefore flows out.
3. Surface tension prevents the water streams from separating again. Surface tension is a phenomenon by which the molecules at the surface of a liquid are held tightly by the cohesive forces of the molecules beneath. Water has quite a high surface tension.

The bucket

The speed of the flow depends on how far below the water's surface the hole is located—or its depth. The depth is the same for both outlets, so water will come out of both holes at the same speed. There's a simple intuitive proof of this problem. Imagine that the two outlets were joined by a piece of pipe.

If water flowed faster through either opening, it would overwhelm the water in the other opening, pushing it back into the bucket and creating a perpetual motion machine.

The mystery of weight

1. The weighing machine would indicate that our weight was decreasing as the Earth's speed of rotation increased.
2. If the Earth's rotational speed continued to increase, then at a certain critical speed, the centripetal force necessary to keep us moving in a circle would be just equal to the total gravitational force. No resultant

force would be left over to provide weight and so the weighing machine would read zero. In other words, we would have become weightless, even though the full gravitational force still continued to act.

How do you build a bridge of bricks?

1. The arched design can accommodate a heavier burden.
2. The amount of material needed is the same for both designs.

Centre of gravity

You drill a hole, H, in your template and suspend it so that it hangs vertically and can swing freely. The centre of gravity is clearly somewhere on the vertical line leading from H directly to the floor. This line can be marked on the template. This procedure is then repeated with a second hole, G, again making sure the line is marked. The point of intersection between the two marked lines is the centre of gravity.

How do you make a cork float centrally?

Pour a bit more water into the glass, not only until it is full, but until the surface of the water is slightly convex above the rim (thanks to surface tension). The cork will naturally float to the highest point of the water dome – the centre – and remain there.

The Archimedes puzzle

Specific weight is the ratio of the weight of a solid to the weight (mass) of an equal volume of water (at 4 degrees Centigrade).

Archimedes could weigh the crown, but his problem was to find the volume of an object as complicated in design as a crown. In a flash ("Eureka! I have found it") he realized that if he immersed the crown in water the volume of water displaced was equal to the volume of the immersed object. Comparing the weight of the crown with the weight of the displaced water gave him the specific weight which, if less than that of gold, would indicate that the metal had been debased.

The brick and the dinghy

Throwing a brick into the dinghy. In the water the brick displaces its volume of water; in the dinghy it displaces its weight of water. Since a brick is heavier than water, it weighs more than its corresponding volume of water does.

Why doesn't the atmosphere crush me?

We do not feel the enormous load exerted on us because our blood pressure balances the atmospheric pressure.

Up in the Air

Why does a golf ball have dimples?

In fact, dimpled golf balls travel about four times farther.

The effect is caused by the backspin imparted to the golf ball by the club. As the top of the ball spins backwards, it drags air which would otherwise have travelled beneath the ball. This air has to speed up to travel the extra distance, and this causes lift in a very similar way to an aircraft wing. If the ball were smooth, the spin would have no effect.

Why don't the clouds move?

As the wind flowed over the mountain, it rose and then descended. This wave continued for a few oscillations over the plain.

The conditions were such that, at the top of the wave, the air cooled enough for the water vapour to condense, forming a cloud at that point. As the air continued along its path, it sank and warmed sufficiently for the water droplets to evaporate again. This caused a series of clouds which were being continuously created at the back and destroyed at the front. The clouds were therefore stationary although the air was moving.

How do birds fly?

Lift is easily explained; it is the same principle as applied to aircraft, involving the Bernoulli effect (see *Glossary* on page 220).

Look at a typical wing, be it that of a plane or a bird:

The air above the wing has to cover a greater distance and therefore travels faster, reducing the pressure. There is greater pressure below the wing; both effects produce lift. This enables the bird to glide and soar through the air.

It is more difficult to explain what provides the forward propulsion. High-speed photography shows that a bird's wing changes shape, flexing in a complex fashion, providing a backward thrust to the air during part of its cycle. As a reaction to this the bird is pushed forward, in a way analogous to a swimmer using the breast stroke.

Certainly a straightforward flapping would be ineffective. Neither do I accept Storer's explanation that the twirling of birds' feathers acts as a propeller.

Why don't planes fly lower?

Even taking the extra distance into account, it is still cheaper to fly at a higher altitude where the aircraft is flying above the turbulent weather patterns. Also, the thinner air presents less drag, meaning less fuel is consumed. Finally, it may be possible to take advantage of the fast winds found at that higher altitude.

Why do boomerangs come back?

The return boomerang has a length of 30 to 75 centimetres, curving to the left, and capable of greater than 90-metre throws. There have been several attempts at explanations. T.L. Mitchell in 1846 suggested that it was caused by the skew combined with the spinning motion. This does no more than beg the question. A more convincing explanation is offered by E Hess in "The Aerodynamics of Boomerangs" (*Scientific American*, Nov 1968).

The boomerang is an aerofoil and therefore subject to a lift, which is greater on the top half because it is turning in the same direction as the boomerang, whereas the bottom half is turning in the opposite direction.

Does the balloon move?

As the car accelerates forward, the balloon on its string tilts forward, too. Inertial forces push backwards in the car, pressing the people against their seats (an effect with which we are all familiar), but also compressing the air at the back of the closed car. This increased air pressure at the rear pushes the balloon forward. For similar reasons, as the car rounds a curve, the balloon tilts into the curve.

Can you fly a jet-powered helicopter?

As the helicopter moves forward and the rotor blades rotate, there is a difference in the lift provided by the forward-moving and backward-moving rotors. This is because the velocity of the aircraft adds to the forward-moving rotors and subtracts from the backward-moving rotors. The faster the helicopter moves, the stronger the effect becomes. The blades flap to compensate, but at very high speeds, compensation becomes impossible and the blades stall, crashing the machine.

Increasing the speed of rotation of the rotors would help, but there is a limit to this. The tip of a rotor is travelling the fastest, and if it should break the sound barrier locally, the resultant disruption to the airflow could cause the helicopter to crash. Also, if the aircraft ever moved forward faster than the backward-spinning rotor blade could retreat, the blade would again stall.

The flag and the weather vane

No flag is perfectly smooth. The tiniest imperfection will make the airflow speed up as it crosses the ripple, because of the Bernoulli Effect (*see Glossary* on page 220). This reduces air pressure on the ripple side, increasing it on the other side, and making the flag flap.

Should I have a fear of flying?

Michael Faraday discovered that there is never any electrical charge inside a hollow conductor; this is called the Faraday Cage Effect. This is why car radios often fade when driving underneath power cables, or steel bridges.

In fact, planes get struck by lightning quite frequently, and the charge passes harmless round the surface of the aircraft. Many passengers would probably not even be aware of the incident. Pilots do try to avoid thunderclouds, however, not because of the lightning, but because of the ferocious up-and-down-draughts that can occur inside them. This could cause structural damage to the plane and interfere with the flow of air over the wings.

One is often given the advice not to shelter under trees during an electrical storm. Sitting in a car would be much safer.

Two cannonballs

Surprisingly, no matter how far apart the cannons are, or at what angle they aim at each other, the balls will always collide in flight. If there were no gravity, the missiles would meet midway between the two cannons. With gravity, the balls fall below this point by equal distances; so they will still collide in mid-air. This is a special case of the fixed-point theorem in topology.

A bolt from the blue

When the lightning bolt strikes a tree, the electricity continues into the ground and causes a voltage gradient across its surface. The nearer the tree, the higher the voltage. Because the cow's front legs are more than a metre from its back legs, there can be a large voltage difference which drives a high, and in this case fatal, current through the poor beast. Because humans have only two feet close together, there was not enough voltage difference to cause the farmer too much trouble.

Can you blow up a balloon?

It is a surprising scientific fact that, as a balloon gets larger, its internal pressure drops. It will therefore be more difficult to expand the smaller balloon. A very simple experiment can demonstrate the fact. Try gently blowing up two identical balloons simultaneously, one partially pre-inflated. You will find that the larger one will inflate in preference to the smaller one.

How do you float a balloon?

Attach an overlong piece of string to the balloon and hold the free end. The string will then form a curve, travelling down from the hand to a low point and then rising up to the

balloon. Cut the string at its lowest point. The balloon will then be left with exactly enough string to cause it to hover.

Dropping a bombshell

When a bomb is released, it has two motions. It starts to fall vertically downwards. It is also moving forwards, initially at the same speed as the plane, although friction gradually slows it down.

If the plane is flying at a low level, there is a danger that the bomb is still directly beneath the plane when it strikes the ground and explodes. This could damage the plane. The function of the parachute is to rapidly retard the horizontal velocity of the bomb, so there is less danger of it inadvertently damaging the low-level bomber.

The stalling plane

A plane can stall when it is travelling too slowly for its wings to provide sufficient lift. When that happens it starts to fall. If, as with most planes, there is more weight at the front, it will fall nose first. As it picks up speed, air flows over the wings, once more providing lift. Eventually there will be enough lift to enable the pilot to pull out of the dive.

This same principle also accounts for the characteristic looping flight that you can see with a paper airplane.

Free fall

The cork will still be suspended in the middle of the bucket because, in free fall, buoyancy ceases to exist.

How do you make a truck fly?

If the bird is in a completely airtight box, the weight of the box and the bird will be the same whether the bird is flying or perching. If the bird is flying, its weight is borne by the air pressure on its wings; but this pressure is then transmitted by the air to the floor of

the box. If the bird is flying in an open cage, part of the increase in pressure on the air is transmitted to the floor of the cage, but part is transmitted to the atmosphere outside the cage. Hence the cage with the bird will be lighter if the bird is flying.

How does a helicopter fly?

Every revolving body develops a force acting perpendicular to the axis of revolution. This moment of force is referred to as torque, and would cause the helicopter to rotate out of control. The tail rotor is designed to counteract the torque.

In large helicopters with two rotors, the torque is avoided by having the rotors rotate in opposite directions.

In the Lab

How can I tell if an egg is raw?

If you spin the egg it will stand on end if it is hard-boiled, similar to a top. A raw egg is unstable because it is asymmetric and the contents are viscous and therefore will not spin.

How do you make salad dressing?

Since oil floats on vinegar, to pour mostly oil Jill had only to uncork and tip the bottle, then cork it, turn it upside down and loosen the cork just sufficiently to allow the desired amount of vinegar to dribble out.

How do I bisect an object?

For symmetrical objects the bisectors will be coincident, but this is not true for asymmetrical objects.

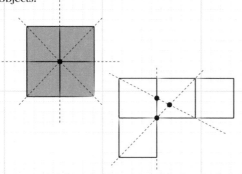

How do you tell acid from alkali?

A red rose petal. Press it hard against some paper, and the resultant pink stain will have the same properties as litmus. In fact a rose blossom can be made a deep blue with ammonia.

Can you blow up a room?

Nothing would happen because there is no oxygen in the room.

Whisky and soda?

There is the same amount of whisky in the soda as there is soda in the whisky. This classic puzzle can be attacked with algebra, fractions, and formulas, or with intuition.

Both containers are left with a litre of liquid after the transfer, the same volume they had to start with. The amount of whisky missing from its bottle is exactly matched by enough soda to bring the bottle's volume back up to a litre. Likewise, the amount of soda removed from its carton, whatever it may be, is now taken up by an equal volume of whisky.

This is true of whisky and soda, but not with all pairs of liquids, such as if a spoonful

of one liquid plus a spoonful of the other becomes less (or more) than two spoonsful of mixture due to chemical reaction.

How do you get an egg into a bottle?

If the unpeeled egg is soaked overnight in vinegar, or acetic acid, its shell becomes plastic. The procedure described will put the egg in the bottle, and a cold-water rinse will restore the shell to its original hardness.

Perpetual motion I

If the magnet were strong enough to draw the ball up the slope, it would be powerful enough to make it jump across the hole. The ball would therefore stick to the flange or magnet.

Throughout history there have been attempts to create a perpetual motion machine: something that would run forever, producing energy. It was not until people had an understanding of the law of conservation of energy that they realized that it was an impossible quest.

Energy can exist in many forms (heat, light, chemical, electrical, potential, kinetic, etc.). The law of conservation of energy says that energy cannot be created or destroyed; all you can do is convert one type into another (e.g., a car converts chemical energy into kinetic, heat, electrical energy, and so on). There is as much energy in the universe now as there was millions of years ago. This law is so fundamental that the British Patent Office refuses any application based on a device which, once in motion, will produce external work indefinitely without any supply of energy.

Although perpetual motion can be observed on a sub-atomic level, for instance electrons orbiting a nucleus, no energy can be drawn from such a system. A perpetual motion machine is, sadly, therefore impossible.

Tables

A three-legged table is more stable on uneven ground, as three points always lie in a plane.

Make a rainbow

The air pressure inside the bubble is greater than the atmospheric pressure outside. It could not be less or it would collapse.

The surface tension necessary to form the bubble is directed towards the centre, trying to reduce the size, and therefore works in tandem with the atmosphere. The inside pressure must, therefore, be greater to balance the two forces acting against it.

The drinking bird

The construction of this gadget is as complex as it is ingenious.

The base is partly filled with a suitable vapour-forming liquid (L) and connected to the head by the tube (T). As T is immersed in L, it produces vapour in the base as well as in the tube and the head. Because the immersion of T in L forms a seal, the two vapour systems are not connected. One should also mention that the centre of gravity (COG) of the contraption is below the fulcrum, so that its position is, to begin with, stable.

Now you wet the top of the bird. As water evaporates, the vapour in the head cools, lowering the pressure. The higher pressure of vapour in the base forces liquid up the tube, relocating the COG until the bird's head tilts forward, dunking its head in the water. In doing so, T is lifted out of L and the two pockets of vapour connect in the tube and equalize pressure. This, together with the buoyancy created by immersing the head, moves the COG to below the fulcrum. The bird rights itself and the whole cycle begins again.

What makes this toy interesting is considering where the energy comes from to keep it going. The answer is that the energy comes from the heat in the water – which gets progressively colder.

We are used to the idea that useful energy comes in very concentrated forms, for example gas or coal. It is unusual to find a machine that can exploit low-grade energy. If perfected, such a machine could be used to power ships, which would then not need to carry fuel. Seawater could be pumped on board, its heat energy extracted to propel the ship, and any resulting cold water/ice dumped back into the sea.

The leaning tower of dominoes

Yes, it's possible. In fact, the top domino can be made to overhang as far as you want beyond the bottom domino without toppling the structure. Success in building such a tower depends on working down from the top; you have to know how many dominoes will be in the final tower before you start stacking. The first stacked domino extends half a domino length. The second projects $\frac{1}{4}$; the third, $\frac{1}{6}$; the fourth, $\frac{1}{8}$; and so on.

After only five dominoes are placed, the top domino (A) is already offset by more than its own length over the fifth domino. With an unlimited supply of dominoes, the offset is the limit of $\frac{1}{2} + \frac{1}{4} + \frac{1}{6} + \frac{1}{8} + \frac{1}{10}$... It is a series that can grow as large as any finite number. On the fiftieth term, the overhang is about $2\frac{1}{4}$ lengths. For three lengths you'll need 227 dominoes, and for 10 lengths you need a whopping 1.5×10^{44}.

Tumbling polyhedrons

No. If you could make a polyhedron that was unstable on every face, you would have created a perpetual motion machine.

Soap bubbles

The small bubble has more internal pressure than the larger one, so it will shrink as the

larger bubble expands. The same principle explains why a balloon is initially hard to inflate, but becomes easier as it expands.

The spinning top

This inversion isn't easy to explain, but various theories have appeared. Jearl Walker, in *Scientific American*, says that the simplest explanation is that the flip-over "arises from friction between the top and the surface on which it spins." That is, whenever the spin axis tilts away from the vertical, the top slides on part of its spherical surface. The friction creates a torque that precesses the top to an inversion. Even with that explanation, this simple toy nevertheless seems to violate common sense.

The Tippe Top raises a second paradox that isn't often discussed. If the top is spinning clockwise at first, as seen from the stem side (left), it will spin anti-clockwise, as seen from the stem side (right), after the inversion.

If you could continuously view the flat part of the stem, the top's rotation would be seen to slow down and stop at some point during the inversion and then to start up again in the other direction! It may seem amazing, but it's true.

The mercury experiment

1. The mercury level will drop to a point, A, such that the pressure of the atmosphere of about 14.7 pounds per square inch exerted on the surface is balanced by the weight of the mercury column (approximately 30 inches, fluctuating with air pressure).
2. This phenomenon was first used by Evangelista Torricelli in 1643 to construct a barometer. Theoretically any liquid could be used for this purpose, but since mercury is about 13.6 times heavier than an equal volume of water, it follows that the water column would have to be about 13.6 times as high as the mercury column, that is, 34 feet of water corresponding to 30 inches of mercury.

Torricelli explained that the column of mercury was supported in the tube by the atmospheric pressure acting on the surface of the mercury in the dish, and pointed out that small changes in the height of the column, which are noticed from day to day, are due to variations in the atmospheric pressure. The space above the mercury in the tube is called a "Torricellian vacuum"; it contains a little mercury vapour, and in this respect differs from a true vacuum.

Torricelli died a few years after the barometer experiment had been performed, and did not live to see his explanation of it, in terms of atmospheric pressure, generally accepted among scientists.

Tongue-in-Cheek Physics

Perpetual motion 2

In theory the system would work; the change in weight, however, is very small (about one-thousandth of one percent), so the amount of energy is minute. Still, this is where the tide's energy comes from.

The energy comes from the rotational energy of the Moon. The tides have the effect of increasing the Moon's orbital radius. This is not, therefore, a perpetual motion machine.

The light and the shadow

The top of the shadow moves faster than the man.

As proof, let A be the position of the man at one point in time, and B be the man's position after he has walked 20 metres. Let AA and BB be the top of the shadows in the two positions. The distance AA–BB is clearly greater than A–B so, consequently, the top of the shadow must have moved faster.

Feathers and gold I

A pound of feathers weighs more. This is because feathers are weighed using the Avoirdupois system, whereas gold (as are silver and drugs) is weighed using the Troy system. A pound on the Avoirdupois scale is greater than a Troy pound.

How heavy can you get?

Yes and no. Our weight is conveyed by the gravitational attraction of the Earth. However, part of this force is counterbalanced by the centripetal force due to the rotation of the globe. This force is zero at the poles, increasing to a maximum at the equator. A person standing, for example, in Quito will have his centre of gravity move up as he grows, thereby increasing the centripetal force, while at the same time reducing the

effect of gravity. Consequently, a point will come when the resultant weight will start to decrease. When the centre of gravity is about 22,000 miles (36,000km) high, the person would be weightless and efficiently in orbit. Needless to say, at one of the poles, the weight would keep on increasing, irrespective of height.

Can you float a battleship in a bathtub?

Strange as it seems, it is perfectly possible to float a battleship in a bathtub. As long as there is enough water to surround the ship completely, it will float. The ship's hull can't 'tell' whether it is surrounded by an expanse of ocean or by a mere fraction of an inch of water. The water pressure on the hull is the same in either case. Likewise, the hydrostatic pressure of the ship is independent of the amount of water that is below or to the side of the ship.

Many people find this hard to believe because they confuse the amount of water 'displaced' with the amount necessary to float the ship. Look at it this way: Suppose the ship weighs 30,000 tons. If the tub is filled to the rim with several million gallons of water and you lower the ship in, the amount

of water that will spill over the sides will, indeed, weigh 30,000 tons. But that's water over the rim. The amount of water left inside, floating the ship, could be a great deal less. If the "fit" between the ship's hull and the tub's wall is tight enough, the water could be squeezed into a thin envelope only an inch thick or less, completely surrounding the ship. This principle is put into practice at Mount Palomar Observatory, where the giant 530-ton horseshoe telescope actually floats in a basin on a thin cushion of oil.

The spying game

No, the operation would not work theoretically. A minute variation in temperature would completely demolish the code, due to thermal expansion.

You could cope with the problem of thermal expansion by converting the decimal into a fraction and cutting the bar into two pieces, one to represent the numerator and the other the denominator. This would be purely a mathematical solution, inasmuch as thermal expansion would affect both bars, leaving the ratio, and therefore the fraction, unaffected.

On a physical level it would not even work theoretically. The problem is the tremendously small distances that one is

soon involved in. Every letter added to the message would involve a hundredfold increase in accuracy. Sending messages with only two letters (e.g., OK) would probably be feasible, for this would require an accuracy of one-tenth of a millimetre.

To send a five-letter message (e.g. ATOMS) would require an accuracy of one-tenth of a millimetre, which is about half the diameter of an atom of iron. As you cannot have half an atom, the sending of such a message by this method is impossible.

Some readers might think that lengthening the bar would provide a viable solution. Ignoring the logistic problem connected with sending a long bar, let us re-examine the position.

Remember that to send the message ISOTOPE MASS, we have to transmit the number 09191520151605391301919. Let us suppose we send a bar whose length is this number of millimetres. How long is the bar?

91,915,201,516,053,913,011,919 millimetres = 10^{23} mm = 10^{17} km*
* 600 million times farther than the distance (93 million miles) between the Earth and the sun.

Air Mail?

In theory the scheme would work, although I doubt that any post office would pay you if they had to deliver a floating parcel.

It is commonly stated that spring balances measure weight (i.e., the force of gravity) and that beam balances measure mass (i.e., the amount of substance in an object). This is not strictly true; the beam balance compares weights, and therefore would not be affected by a change in gravity. For example, kitchen scales would give the same reading on the moon, but a spring balance would not

In this case, however, even a beam balance would not give a correct reading of the mass of the parcel. I do not think that there is a common instrument that would. Either the parcel should be weighed in a vacuum or a machine would have to be invented that measured the inertial mass by, perhaps, gently oscillating the parcel.

The ultimate weapon

It would not work, even theoretically. In the first instance the major devastation would occur in the country resorting to such a weapon although, according to David Stone's theory, there would be damage in the U.S. provided the jumps were timed in regular intervals to amplify the resulting ground waves. Stone estimates, however, that a jump would have to take place about once every hour for some considerable time. So, apart from the devastation at home, the attacking country's economy would certainly suffer irreparable damage.

Theoretically at least, the U.S. might indeed consider inviting all NATO members to organize counter-jumps, carefully timed, to cancel the ground waves initiated by their adversary as an alternative to using advanced weaponry in defence of their shores and avoid the possibility of starting a world conflagration. Before they do so, however, let us now do a simple calculation to work out how much

sway and some damage locally. Put another way, this is about one four-hundreth of the amount of energy released by the atomic bomb over Hiroshima.

The theory goes that the waves of energy from the jump would spread around the world and focus on a point directly opposite. However, the Earth would rapidly dissipate the energy, converting it into heat. We know this because, when an earthquake occurs at one point in the Earth, it does not cause damage at its antipodes. So the U.S. and NATO can stand down their anti-pogo defences.

Charging for a flight?

No. It is true that a voltage would be generated between the wing-tips as the plane flew through the Earth's magnetic field. However, to make use of it we would have to complete the circuit, and that would require running wires to the ends of the wings. Those wires would also be travelling through the Earth's field and would also generate the same voltage. It would be like connecting two identical batteries back to back: the two effects would cancel each other out.

If we consider other forms of transport, we might be more successful. By a similar process a train would generate a voltage which could be picked up from the rails by contacts lying beside the track. I calculated that a train travelling at 100 mph (160 kmph) would generate about 1.5 millivolts. That is obviously not enough to solve any energy crisis! NASA is experimenting to see if this is

energy would be produced by each attacking jump, and the extent of the danger to the U.S.

No. of Chinese people	= one billion
Average mass of 1 person	= 50 kg
Pogoing height	= 0.5 meters
Acceleration due to gravity	= 10 m/s²

Calculation:

Energy produced:
$$= 1,000,000,000 \times 50 \times 10 \times 0.5$$
$$= 2.5 \times 10^{11} \text{ joules of energy}$$

This would produce a local earthquake of about 5.3 on the Richter scale, causing trees to

a viable power source for spacecraft, however. A long wire, several miles long is trailed through the Earth's magnetic field. The electric charge is dissipated from sharp points at both ends of the wire.

How do you fly through time?

Yes, there certainly is. First, let us distinguish between natural time and an arbitrary time zone. A time zone is an area within which everyone has agreed to use the same time. However, every point on the Earth's surface has its own natural time. As one travels east, the natural time is farther ahead, one hour for every 15 degrees of longitude.

Let us suppose that the aircraft travels eastwards around the world so fast that it would complete the journey in one second, and it started from a point just to the east of the International Date Line. As it travelled it would encounter a natural time which was progressively further ahead.

When it arrived at a point just to the west of the International Date Line the time would be 24 hours ahead. This extra day is then lost as the date line is crossed. The final time is then exactly one second after the start.

One cannot, therefore, travel into the past or future by using the International Date Line.

Feathers and gold 2

An ounce of gold weighs more. The Troy system has 12 ounces to a Troy pound, whereas in the Avoirdupois system a pound

consists of 16 ounces. So a Troy ounce is greater than an Avoirdupois ounce.

The two Feathers and Gold puzzles demonstrate the vital importance of defining precisely the units in which a measurement is made. Units of measurement are objective, ultimately reflecting some measure in the world as observed by man. Thus, the measure of length is a foot, the measure of the size of a horse is the hand. Both the Troy system and the Avoirdupois system are based on the weight of a grain of wheat as developed by man, which (to use another system, metrics) is 0.0648 grams. A Troy pound is 5,760 grains, and an Avoirdupois pound is 7,000 grains. In the words of the first Greek sophist, Protagoras (fifth century B.C.): "Man is the measure of all things." Whereas Protagoras meant this subjectively, to suggest that judgments are relative, it is better interpreted objectively, to express the nature of units of measurement.

How do I make a sound bomb?

The design would not work. First, there is no such thing as a perfect reflector of sound. Second, the amount of energy in a watch is quite small. But even without these two objections the proposal would still be impractical.

Air is a very inefficient transmitter of sound and consequently sound energy is continually being converted into heat. The energy thus escapes from the box in the form of heat.

Glossary

Assuming that readers may not be familiar with all the scientific terms used in this book, it should be useful, for ease of reference, to define here some concepts not part of everyday vocabulary for your use when solving these puzzles.

absolute zero — the temperature at which the molecules of any object are stationary; this, therefore, represents the coldest possible temperature.

adhesion/cohesion — adhesion is the force of attraction between two different types of molecule. The force of attraction between similar molecules is called cohesion.

adiabatic process — a term used in thermodynamics, referring to a condition in which no heat enters or leaves the system, although pressure and volume are varied. For example, the air in a bicycle pump will heat up if compressed, because no significant heat transfer will take place immediately. Another example, with the reverse effect, is an aerosol can. On releasing the contents, the temperature of the can will drop. Other systems in daily use, such as automobile engines and refrigerators, exhibit adiabatic phenomena.

Archimedes' principle — states that when any object is immersed in a fluid there is an upthrust that acts on the object which equals the weight of the displaced fluid.

Bernoulli effect — discovered and formulated by the Swiss mathematician Daniel Bernoulli in 1738. The principle states that, as the speed of a moving liquid or gas increases, the pressure within that fluid decreases. This principle is an important aspect of aerodynamics and covers the flow over surfaces, such as the wings of aircraft and ship's propellers. As the air flows over the upper surface of a wing it speeds up, and consequently suffers a reduction in pressure as compared to the lower surface. The resulting difference in pressure provides lift to the aircraft.

centre of gravity — the point in any object where all the weight could be considered to be concentrated without affecting the properties of the object. Such an assumption can greatly simplify calculations.

cohesion — See adhesion.

conduction of heat — the process by which heat energy can travel through a substance by the transfer of vibrational energy between adjacent molecules. This is greatest in solids.

convection of heat — the process by which hot fluid is forced to rise by being displaced by colder fluid.

Coriolis effect — a deflection caused by the rotation of the Earth.

diffraction — a process by which a wave curls around small objects or spreads out as a result of passing through a small gap.

drag — See friction.

energy conservation — the scientific law that states that energy (heat, light, sound, electricity, etc.) cannot be created or destroyed, but can only be converted into another form.

evolution — the theory proposed by Charles Darwin that states that species change over long periods of time, as a result of which individuals that possess advantageous differences are more successful in procreation.

fluid — a substance that flows: a liquid or a gas.

friction — a force that results from one object moving relative to another with which it is in contact.

gravity — a force of attraction between any two objects. The force is very small and is only noticeable if at least one of the objects is very large.

LED — stands for Light Emitting Diode, a small solid-state device that produces light, without heat, when a small electrical current is passed through it.

lift — the upward force produced by the flow of air over wings. If the lift disappears because of a disruption in this flow, the wing is said to stall.

light-year — the distance that light travels in a vacuum in one year, about 5.88 trillion miles (or 9,460,000,000,000,000 metres).

mass — measure of how much material there is in an object. This is independent of the force of gravity (weight) acting on the object. Weight is measured by a spring balance and mass by a beam balance.

Newton's laws of motion — (1) An object will remain at rest or in uniform motion unless acted upon by an unbalanced force. (2) When an unbalanced force acts upon an object, it will accelerate at a rate proportional to the force and inversely proportional to the mass of the object. (3) For any action on an object, there will be an equal and opposite reaction.

perpetual motion machine — refers to a mechanical device doing work and operating perpetually without any supply of energy, other than that which is generated by the device itself. Such a system cannot exist, as it is contrary to a well-established physical law, namely the principle of 'conversion of energy.' This law is so fundamental that patent offices have been known to refuse any application based on perpetual motion. Perpetual motion is perfectly feasible in the absence of friction; for example, electrons rotating around a nucleus or an object flying through space. However, as soon as energy is extracted from such a system, it will slow down, destroying its perpetual motion. Needless to say, any device using the force of gravity or variation of atmospheric pressure or temperature would not qualify.

polarization of light — light is a waveform that has an electric and magnetic vector. In all forms of natural light these vectors are not aligned. If aligned, the light is said to be polarized.

polarizing filter — A filter that allows light

to pass which is vibrating in one direction only. See polarization of light.

principle of flotation — states that a floating object displaces its own weight of fluid.

radiation of heat — the process by which heat can travel through a vacuum or a fluid. It is a form of light (infrared).

Rayleigh scattering — named after Lord Rayleigh (1842-1919), a type of deflection of electromagnetic radiation by particles in the matter through which it passes. The radiation photons bounce off the atoms and molecules without any change of energy (elastic scattering), changing phase but not frequency, as opposed to inelastic scattering.

refraction — the process by which the direction of light changes when it travels from one medium to another of different density. This is how lenses work.

stall — See lift.

surface tension — a condition existing on the surface of liquids giving them film-like characteristics. The tension is explained as resulting from intermolecular forces of the liquid. Examples: A water beetle can ride the surface of water; the near-perfect sphere of a soap bubble; a small quantity of mercury poured onto a plane assumes a near-spherical shape, flattened only slightly by gravity.

weight — See mass.